CELESTIAL NAVIGATION

Jeff Toghill

W·W·NORTON & COMPANY
New York *London*

ISBN 0-393-30294-6

W. W. Norton & Company, Inc., 500 Fifth Avenue, New York, NY 10110
W. W. Norton & Company Ltd, 10 Coptic Street, London WC1A 1PU

8 9 0

Contents

Introduction

The success of my previous book 'Coastal Navigation' brought home the remarkable interest of yachtsmen and women, both sail and power, in the subject of navigation. That book was confined to navigation within sight of land, and there was an obvious need for a similar publication dealing with navigation on the high seas.

This book fills that need and I have taken the same approach as I did in 'Coastal Navigation' in simplifying the subject to a suitable level for small craft. The complex subject of 'big ship' navigation is deeply involved and often requires sophisticated equipment. Few yachts carry such equipment and few yachtsmen or women have the time or the interest to study the subject in such depth.

What they need is a practical approach to navigation which enables them to sail their vessels across oceans from departure to arrival points with the minimum of fuss and the minimum of expense, but with maximum safety. This book deals with the subject as it is encountered in practice aboard small craft and is based on my 35 years experience in yacht navigation in all corners of the globe.

Jeff Toghill
1986

Defense Mapping Agency Hydrographic/Topographic Center charts and some illustrative pages from the Center's publications reproduced by courtesy of the Defense Mapping Agency Hydrographic/Topographic Center, Washington DC, USA.

Chapter 1 **Introduction to Celestial Navigation**

Celestial navigation is very similar to coastal navigation.

Both involve the use of an instrument, both are ultimately plotted on a chart, and both use a series of not dissimilar fixes to establish the boat's position. But there are differences: coastal navigation uses shore objects as a means of obtaining a fix, where celestial navigation uses sun, moon, planets or stars. And between the use of the instrument and the plot on the chart, celestial navigation requires a certain amount of calculation and use of tables, where coastal navigation requires only the conversion of compass bearings to true.

The instruments

Both coastal and celestial navigation use Mercator charts and identical chart instruments for plotting. The instrument used for obtaining the sights, however, differs. A compass is used for coastal work, while a sextant is required for celestial sight taking. In both cases, the distance log is a common instrument and is used in an identical manner.

The calculations

Whereas the only calculation required for coastal navigation is simple application of compass error to the bearings taken, celestial sights require a certain amount of calculation, depending on the particular type of sight used. The noon sight, where latitude is obtained, requires only the addition of two lines of figures; whereas a full Marc St Hilaire sight involves the use of a complicated mathematical formula.

This latter form of sight, however, is rarely used these days, and modern sight reduction methods involve no more mathematics than the addition of a few lines of figures, and do not involve the use of trigonometry or formulae. The sight reduction tables act as a sort of computer and from information fed into these tables, a

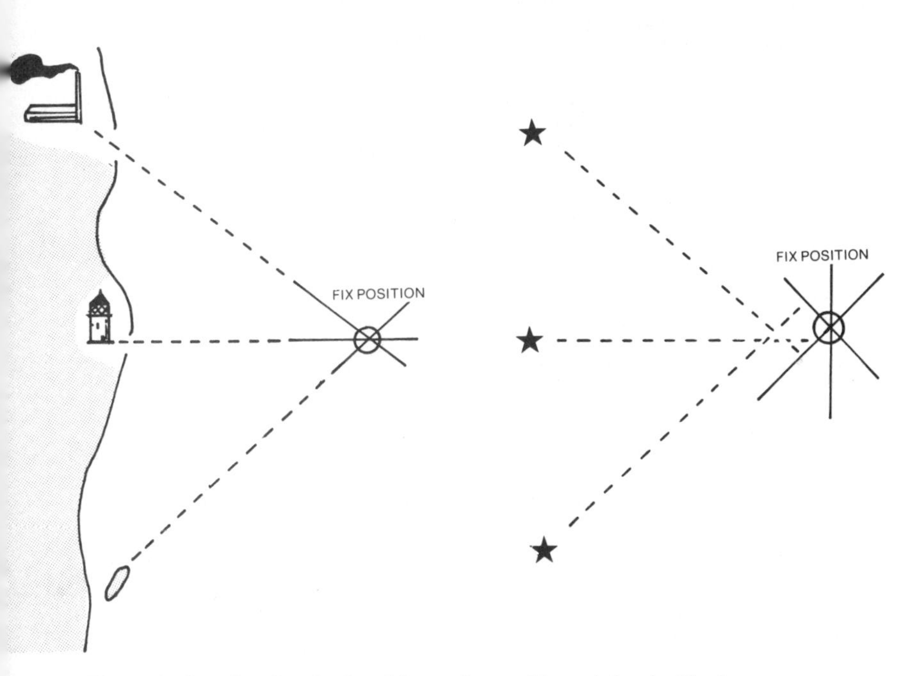

The cross bearing fix obtained from shore objects is basically the same as that obtained from a star sight.

computation is taken out which resolves simply into a plotting position. Thus, where until recent years celestial navigation required quite involved mathematics, the sight reduction method commonly in use today, requires little or no brain strain. The only tables required are those in *The Nautical Almanac* and the *Sight Reduction Tables*.

The methods I have developed for yacht navigation as used in this book simply mean filling out a set form which, as mentioned,

The sextant is the only instrument for celestial sight taking.

LATITUDE BY MERIDIAN ALTITUDE

(NOON SIGHT)

DATE 25-12-85

LMT 1200 GMT 16 h 04 m 06 s

Noon Alt 77° 41'

Dip 3' −

App Alt 77° 38'

Alt Corr 16 +

90 00

TRUE ALT 77° 54'

ZX 12° 06'

Dec 23° 14'S +

NOON LAT 35° 20'S

The noon sight is the shortest computation in celestial navigation.

The Marc St Hilaire computation is the most involved but is rarely used nowadays.

POSITION LINE BY MARC ST.HILAIRE METHOD

Star Arcturus GMT 10 h 01 m 12 s

DR Lat 35° 31'S Long 118° 16'E

GHA Aries 126° 35.9'

Inc 18.0' +

126° 53.9'

SHA star 146° 34.7' +

GHA star 273° 28.6'

DR Long 118° 16.0'E +

LHA 31° 44.6'

Dec 19° 24.5' N

DR Lat 35° 31.0' S +

Diff 54° 55.5'

Sext Alt 27° 30.5'

I.E. 0.0'

Obs Alt 27° 30.5'

Dip 7.2' −

App Alt 27° 23.3'

Alt Corr 1.8 −

90° 00'

True Alt (Ho) 27° 21.5' −

ZX 62° 38.5'

Log Hav (LHA) 8.87382

Log Cos (Lat) 9.91060

Log Cos (Dec) 9.97459

8.75901

Nat Hav 0.05741

Nat Hav (Diff) 0.21268

Nat Hav (ZX) 0.27009

ZX (obs) 62° 38.5'

ZX (calc) 62° 37.5'

INT 1.0 Away

A 1.14 N

B 0.67 N

C 1.81

Az N34°W (326°)

***The Nautical Alamanac* and *Sight Reduction Tables* are the only two volumes required for basic yacht navigation.**

has no more mathematical involvement than simple addition or subtraction—the tables do the mathematical work for you. The sight reduction method, which is used by professional navigators of naval and commercial shipping the world over, is now accepted as being standard procedure for most celestial sight taking.

A greater amount of calculation is required for celestial sight taking than for coastal position fixing. This is because the celestial bodies used for sight taking are constantly moving, travelling around the globe in twenty-four hours, and therefore their position at the time of sight taking must first be found before they can be used to establish the boat's position. Shore objects used for coastal navigation are static and their position is known.

The plot

While both methods involve the use of Mercator charts, it follows that the scale of these charts will be somewhat different. The coastal chart is usually of medium to large scale allowing for

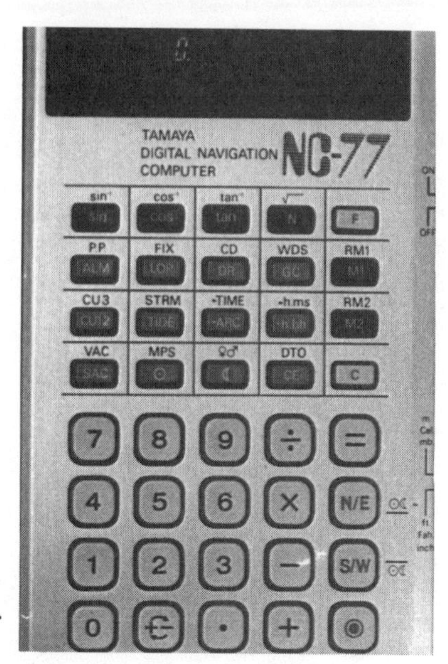

Portable calculators can be programmed for sight reduction calculations.

measurements of one mile or less. The small scale ocean charts, however, rarely permit measurement of less than ten miles, and are thus not suitable for the close work involved in plotting a sight. To overcome this, Plotting Sheets are obtainable from chart agencies, and these are simply a 'blown up' section of the ocean chart, enlarged to the point where plotting over short distances is feasible and accurate.

There are other methods of plotting celestial sights, and many professional navigators prefer to draw up their own plotting sheets or devise their own systems. However these are too numerous to include in a book of this type and in any case they require a deeper understanding of the celestial situation. Plotting sheets are the simplest form for the beginner, and use precisely the same principles as normal coastal chart plotting.

The fixes

The similarity between coastal and celestial navigation is probably best seen in the methods of obtaining a fix of the boat's position. In coastal navigation, when more than one object is visible, a cross bearing fix is obtained. With only one object in sight a running fix is used. Similarly, in celestial navigation, sun sights taken during the daytime when only the sun is visible employ the

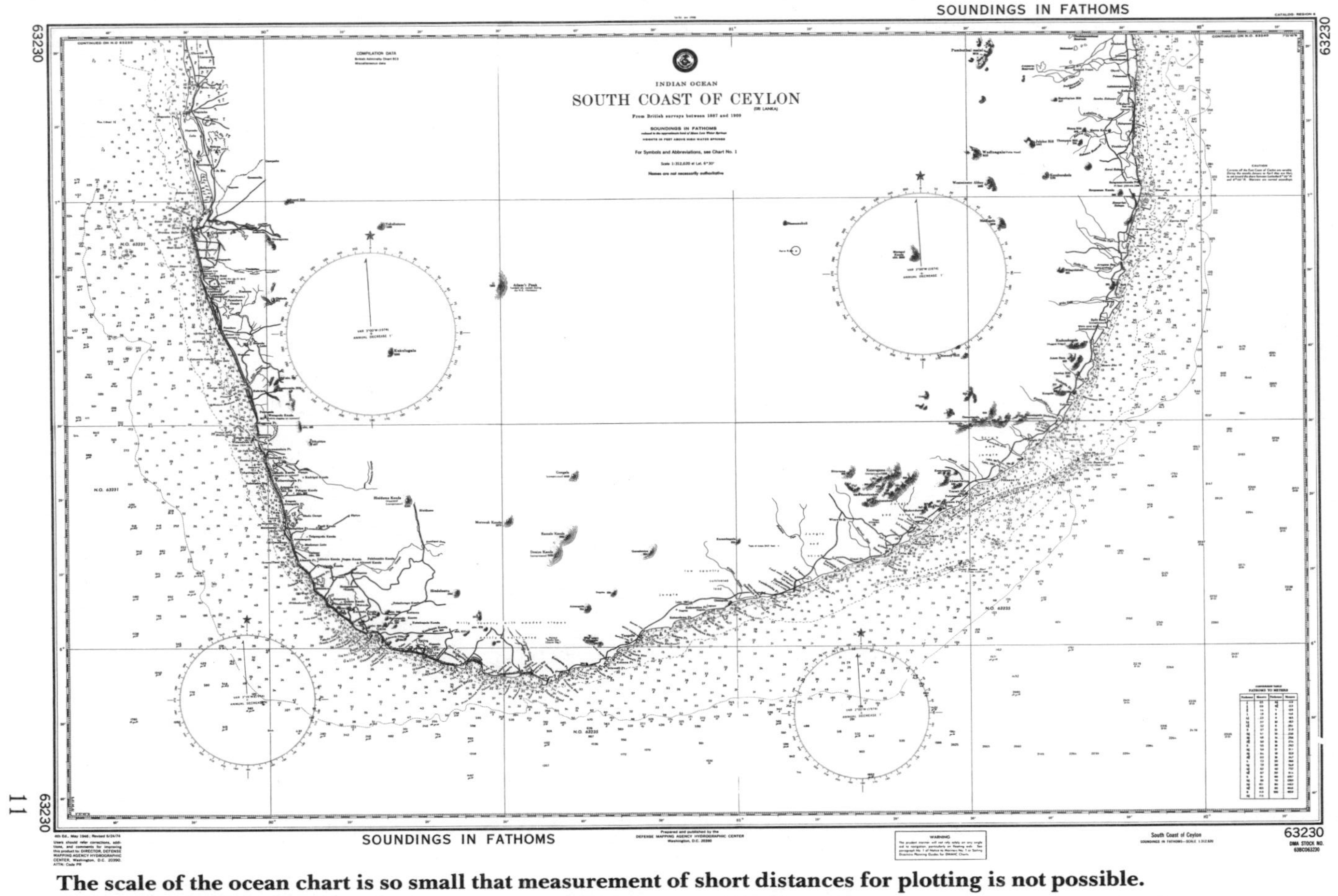

The scale of the ocean chart is so small that measurement of short distances for plotting is not possible.

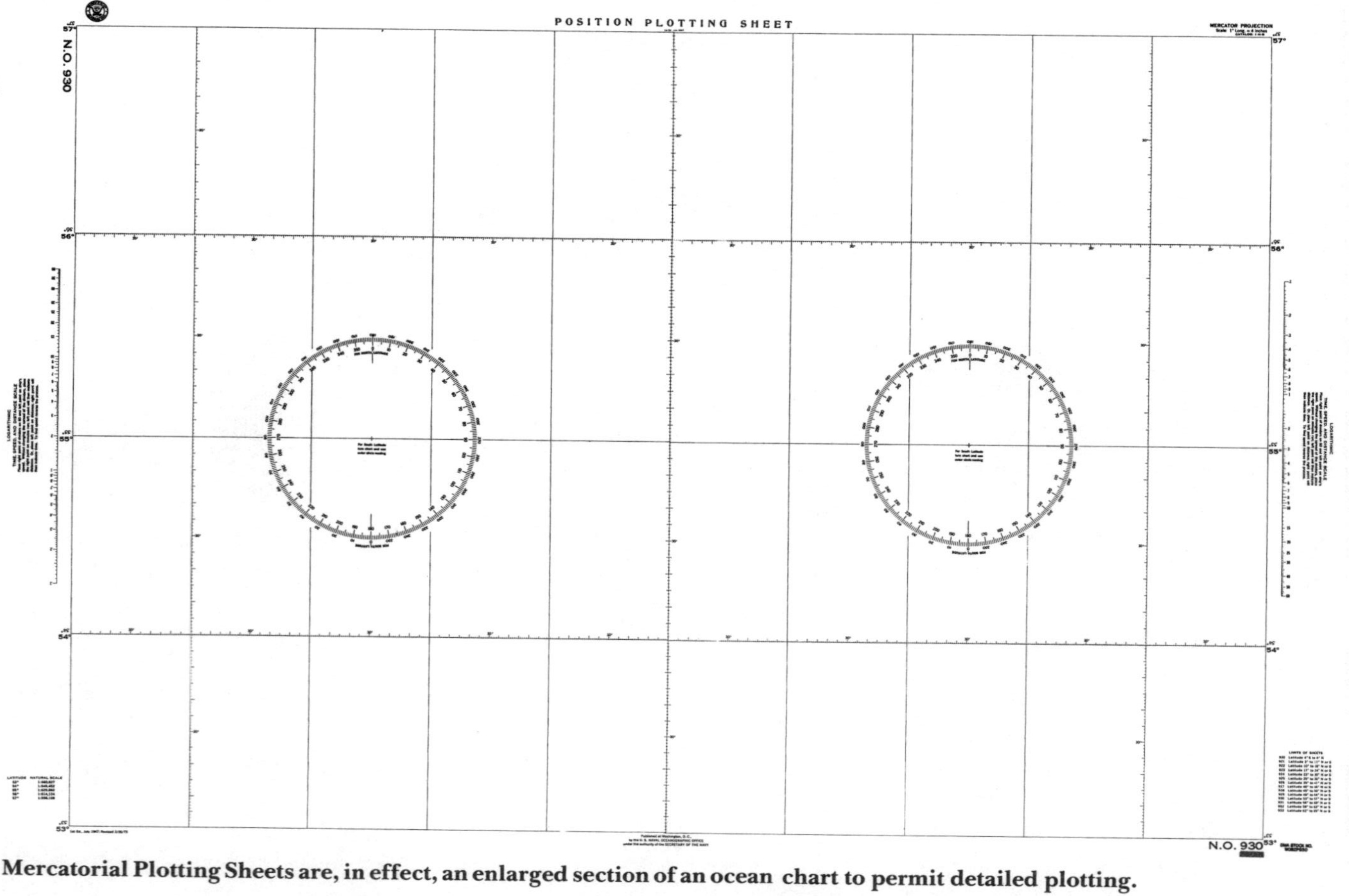

Mercatorial Plotting Sheets are, in effect, an enlarged section of an ocean chart to permit detailed plotting.

running fix method. But a full cross fix can be obtained when using stars and planets because of the number available for use.

The navigator's day

The sextant measures the angle of a heavenly body above the horizon and, therefore, to obtain accurate results not only must the heavenly body be clearly visible, but the horizon must be sharp and clear. For this reason sextant sights are never taken at night. Sun sights are taken during intervals throughout the day and star sights are obtained at dawn and dusk when the sky has darkened sufficiently to pick out the heavenly bodies, but the horizon is still light enough to give a sharp clear line on which to measure them.

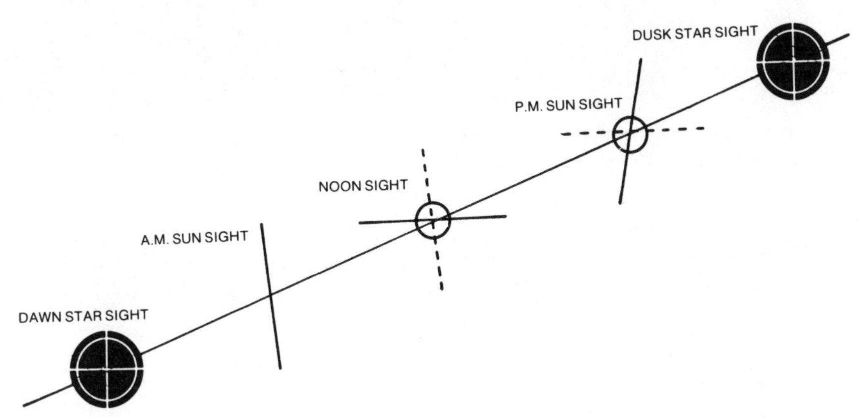

The navigator's day.

In terms of sight taking, then, the navigator's day begins at the crack of dawn and continues until the last light at dusk. A full fix is obtained with both morning and evening star sights, but during the day, unless the moon happens to be visible, running fixes by means of sun sights is the only available method of obtaining a plot. A common routine for the navigator's day would be as follows:

1 *AM stars (Dawn)*
A full fix obtained by using three or more stars or planets (sometimes also the moon).

2 *Morning sun sight*
A single sight of the sun for obtaining a position line. This will also be used in a later running fix.

A clear, sharp horizon is essential for accurate sight taking. For this reason sights are never taken at night.

3 *Noon sun sight*
A latitude by noon sun shot, which can be crossed with the morning position line to obtain a running fix position.

4 *Afternoon sun sight*
An optional sight which produces a position line. This can be crossed with the noon latitude to give a running fix of the boat's position.

5 *PM stars (Dusk)*
A full fix obtained by using three or more stars or planets at dusk.

Geographical position

This is the spot on earth directly beneath a heavenly body. In the case of the sun it would be the spot where your body would throw no shadows. It is better defined as the point at which a line drawn from the heavenly body to the center of the earth cuts the earth's surface.

Altitude

The angle of a heavenly body above the horizon measured by the sextant is called its altitude.

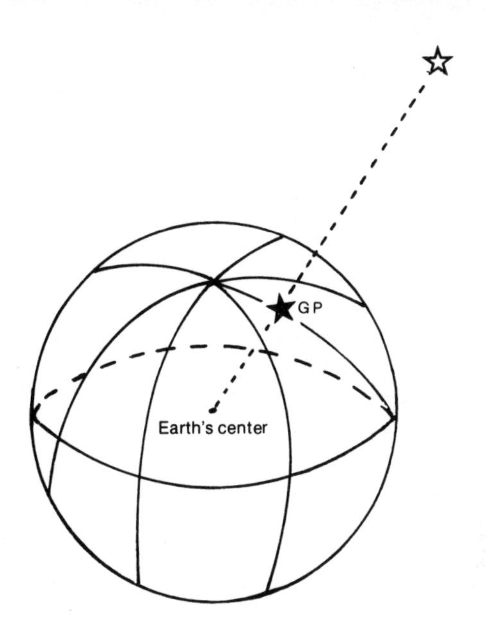

The geographical position of a star.

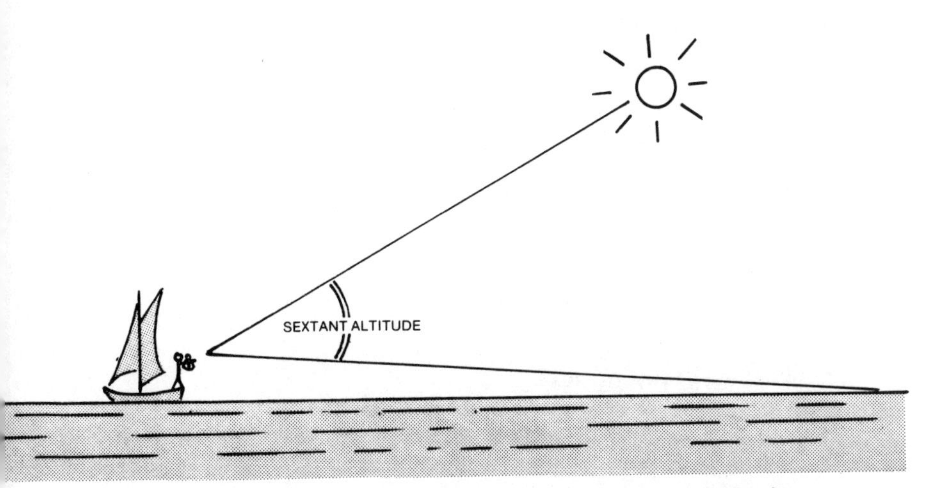

The sextant measures the angle of the sun (or stars) above the horizon.

Circles of equal altitude

In diagrammatic form, an altitude can be illustrated as a right-angled triangle with the heavenly body at the apex. It is obvious that an identical altitude can be taken on the opposite side of the

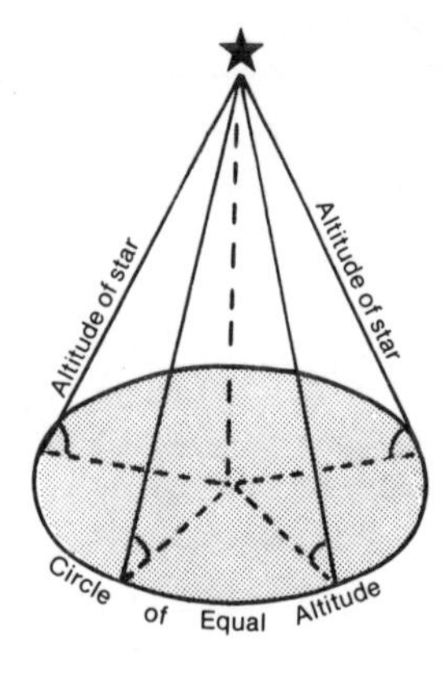

Equal altitudes of the same heavenly body are measured from the circumference of a circle. The boat's position must lie somewhere along this circle.

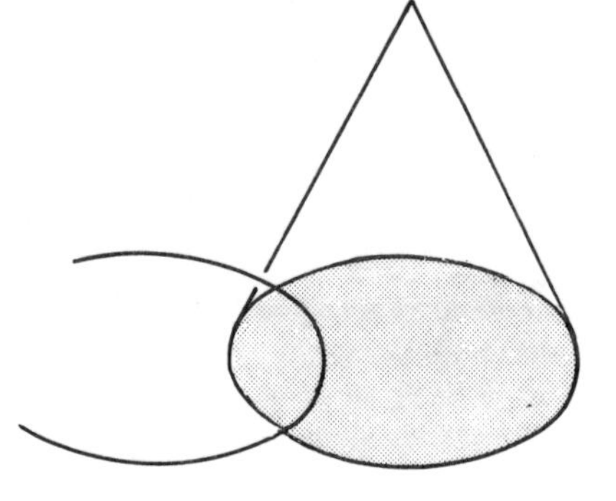

Two circles of equal altitude taken at the same time intersect in two positions. One must be the position of the boat.

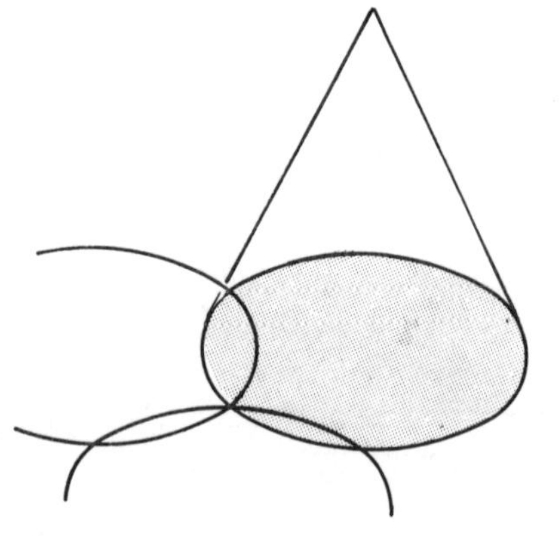

By obtaining a third circle of equal altitude from another heavenly body, the exact location of the boat is found.

heavenly body and in fact, a whole series of equal altitudes can be taken from different positions around the heavenly body. An illustration of this is a cone in which the angle at the base is identical anywhere around its circumference.

The circle formed by the base of the cone is termed a *circle of equal altitude* and the position of the navigator, shooting this altitude on his sextant, must lie somewhere around the circumference of the circle.

The question is, just where along this circle? By shooting another heavenly body, another circle of equal altitude will be obtained and where the two circles intersect must be the navigator's position. The circles intersect at *two* positions but because of the scale of these circles on the earth's surface, the two positions will be many hundreds of miles apart and you should know to within less than a few hundred miles where your boat's position is! But in any case the problem can easily be resolved by taking a sextant sight of a third heavenly body which, like the cross bearing fix of coastal navigation, will intersect with the other two, establishing the boat's position beyond any shadow of a doubt.

Position line

Since a circle of equal altitude may cover an area of the earth's surface many thousands of miles in circumference, it is impossible to plot on a chart. Therefore a tiny section of the circle in the immediate vicinity of the boat is the only part used for plotting. Again because of the size of the circle, this segment of the circumference, when plotted, can be assumed to be a straight line. It is termed a *position line* and the location of the boat lies somewhere along this line.

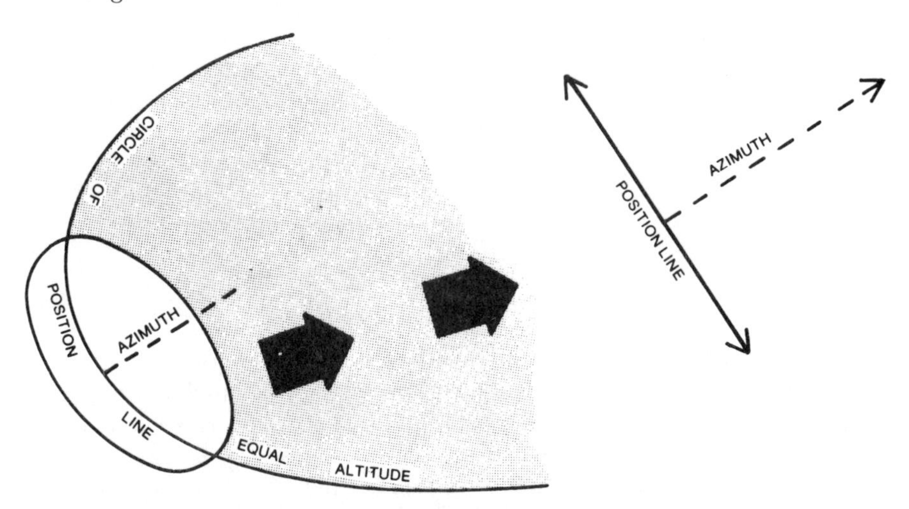

The small portion of the circle of equal altitude in the vicinity of the boat can be represented by a straight line known as the 'position line'.

Azimuth

This is just another name for a bearing of the heavenly body. It can be represented diagrammatically by the radius of the circle of equal altitude. Since the radius of a circle lies at right angles to its circumference, the azimuth of the heavenly body is plotted at right angles to the position line.

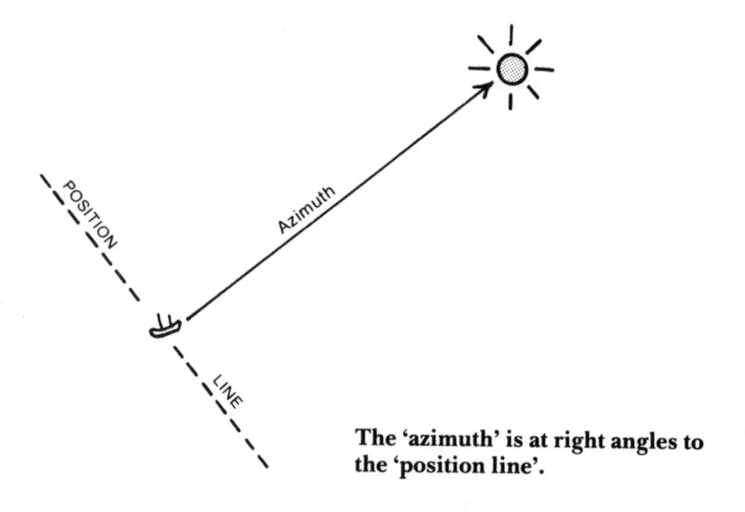

The 'azimuth' is at right angles to the 'position line'.

DR and EP

Dead reckoning (DR) is the same in celestial navigation as in coastal navigation. It is the position of the boat as accurately as can be calculated from records of her past progress. It can be found either by measuring her track across the chart or by calculation (see Chapter 4). An *estimated position* (EP) is a DR corrected for any current or other factors which may push the boat off course. Where possible an EP is used for working celestial sight computations.

The universe

The universe is infinite. Despite the advances of modern technology, space exploration and the giant telescopes which allow us to look into the farther reaches of space, still no limit to the extent of the universe has been found.

The universe is composed of galaxies of different sizes and shapes. Our own galaxy is the Milky Way and is but one of thousands, perhaps millions, which make up the universe.

Andromeda—a typical galaxy. Note the plate-shaped, spiral appearance.

Phases of the moon.

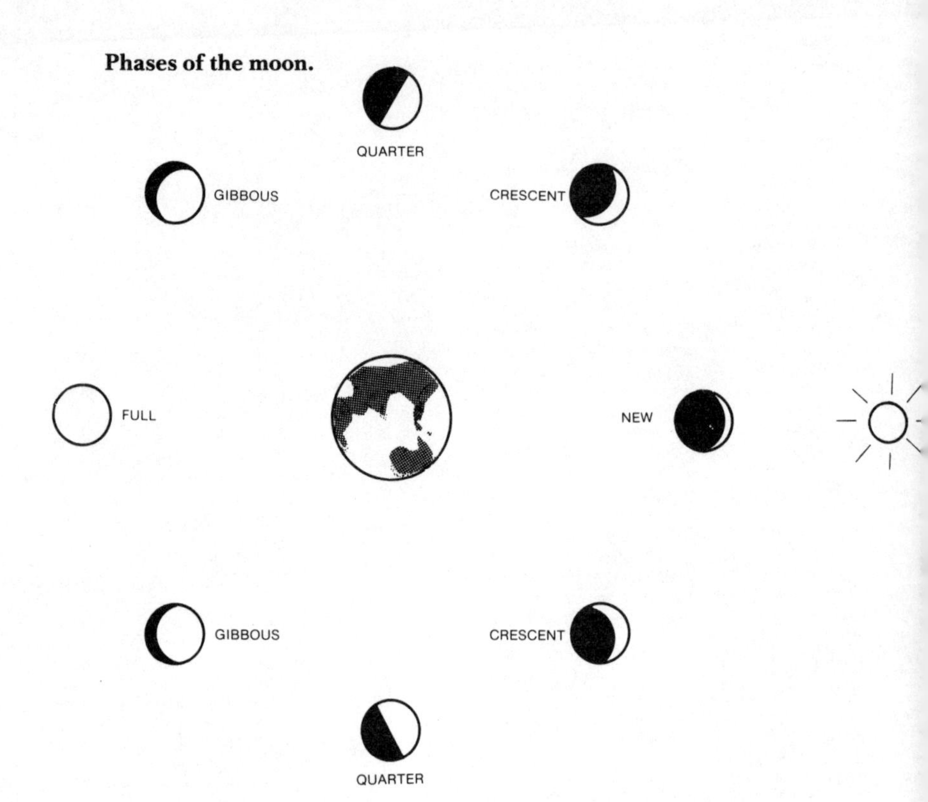

Galaxies

A galaxy is a roughly plate-shaped gathering of stars and similar astronomical units. The number of stars in each galaxy runs into billions and although galaxies may vary in shape and size they are all immense and all roughly resemble the shape of a plate or record—circular and flat. Since earth lies on the outer edge of its galaxy this shape can be seen by looking into the cross-section of our own galaxy. This is the Milky Way, thin in cross-section but spread across a large part of the night sky. The circular shape cannot be seen, but the cross-section is very obviously similar to a plate or lens section. In many galaxies, the circular plate shapc is made up in the form of a spiral with the outer edges spiralling into the center.

Our galaxy is named the Milky Way from the milky appearance of the myriad of stars which make up its cross-section as seen from earth. The largest stars are known as *first magnitude* stars and it is these which are used mostly for navigation. Fifty-eight of these

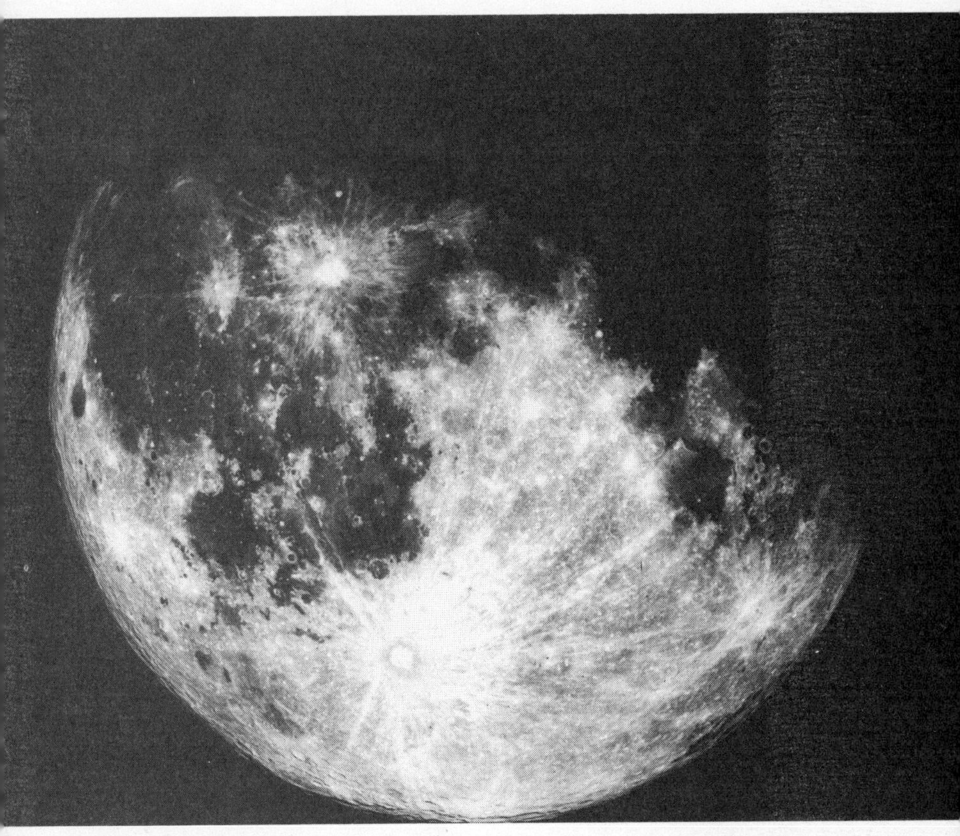

The moon, despite its size and magnificent appearance, is not the most popular object for navigational purposes.

The solar system

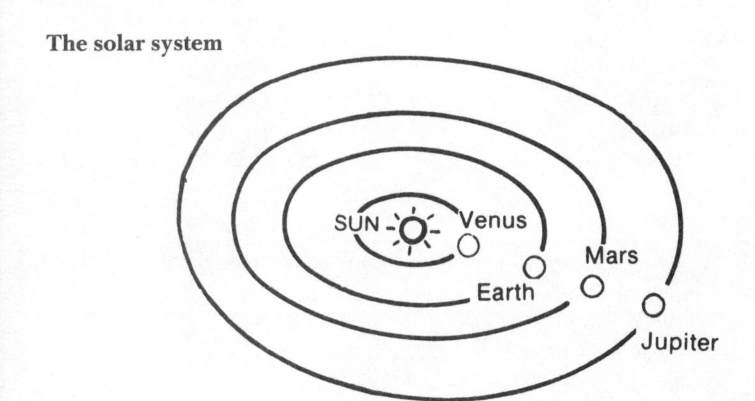

stars are listed in *The Nautical Almanac* as being 'selected stars' which are of a magnitude and location convenient for navigation. Polaris, the Pole Star, is included as one of these stars since it is the most useful star of all for ocean navigation.

Solar system

Our solar system, the center of which is the sun, lies towards the outer edge of our galaxy. It consists of nine major planets and thousands of small planetoids or asteriods travelling constantly in orbits around the sun. Like the stars, only a few of the planets are of use for navigation, those listed in *The Nautical Almanac* being Venus, Mars, Jupiter and Saturn, all of which are easily visible due to their magnitude. Venus, the brightest planet, is closest to earth, while Saturn is farthest away. An idea of the distances involved can be gained from the fact that Venus completes a revolution around the sun in about 225 days while Saturn takes between twenty-nine and thirty years.

The planet Saturn as seen through a 100-inch telescope.

The moon

The moon is earth's only satellite. Its orbit is elliptical, its distance from the earth varying from between 220,000 and 253,000 miles. This explains the apparent change in the size of the moon when observed from earth. In the course of its orbit, the moon also turns on its axis, thus presenting the same face constantly towards earth. This face may be lit at different relative angles, depending on the relation of the moon and the sun in their particular orbits.

The moon passes through a complete cycle in 29½ days. Due to the relative positions of earth, sun and moon, the face of the moon changes throughout this cycle according to the angle at which the sun's rays strike its face. When the moon is between the sun and the earth, for example, its sunlit face cannot be seen from earth and therefore there is no moon at all; *new moon*. As the moon moves eastward from the sun, the sun illuminates an edge which is seen from earth as a 'crescent' moon.

As it moves farther east, more and more of the moon's face is illuminated reaching the point where, in the first and last quarter, half the moon's face is lit. When sun and moon are on opposite sides of the earth, the full face of the moon is illuminated and it is seen completely lit as a *full moon*. The moon then begins to move toward the sun and the lighted face of the moon 'wanes' until finally, with the sun again behind it, the moon completes its cycle.

Meteors and meteorites

These are often called 'shooting stars' and are small bodies in the solar system—often no larger than a grain of sand—which burn out brilliantly as they enter the earth's atmosphere. Most are burned up before they strike the earth's surface, but those that do reach earth sometimes create sizeable craters.

Magnitude

In navigation the usefulness of a heavenly body depends on its brightness, so a system of rating the brightness of stars and planets has been developed and is termed *magnitude*. The magnitude is listed in *The Nautical Almanac*—planets in the daily pages, stars in the list of selected stars. All prominent stars and planets used in navigation fall into the category known as first magnitude. Surprisingly, the smaller the magnitude the brighter the star, most very bright bodies having a negative or minus magnitude.

Chapter 2 The Sextant and Chronometer

In coastal navigation, a fix of the boat's position is obtained by relating it to shore objects whose positions are known. The same procedure is used in celestial navigation using the sun, moon and stars instead of shore objects. In place of the compass the sextant is used and for this reason, the sextant is the most important instrument for celestial navigation. It is important to understand the techniques and the problems involved in using it.

Principle

The sextant works on the same 'split-image' principle as the camera rangefinder. By means of mirrors, two separate objects or parts of objects are brought together in the eyepiece and a measurement recorded. In the case of the rangefinder this measurement is distance, in the case of the sextant it is the angle between the two objects. Thus, by using the sextant vertically and the sun as one object and the horizon as another, the angle of the sun above the horizon is found when the two are brought together in the telescope of the sextant.

Using mirrors, one of which is fixed and one adjustable, the double-reflecting principle of the sextant permits great accuracy. Measurements of angles up to about 120° can be made although the arc of the sextant itself covers only 60°.

Sextants come in all shapes and sizes. This one, designed for yacht use, is compact and accurate.

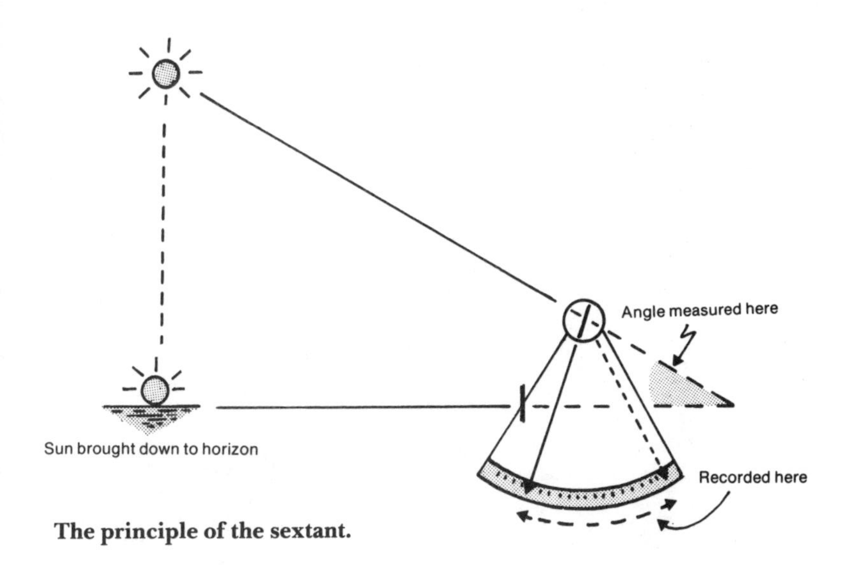

The principle of the sextant.

Parts of the sextant.

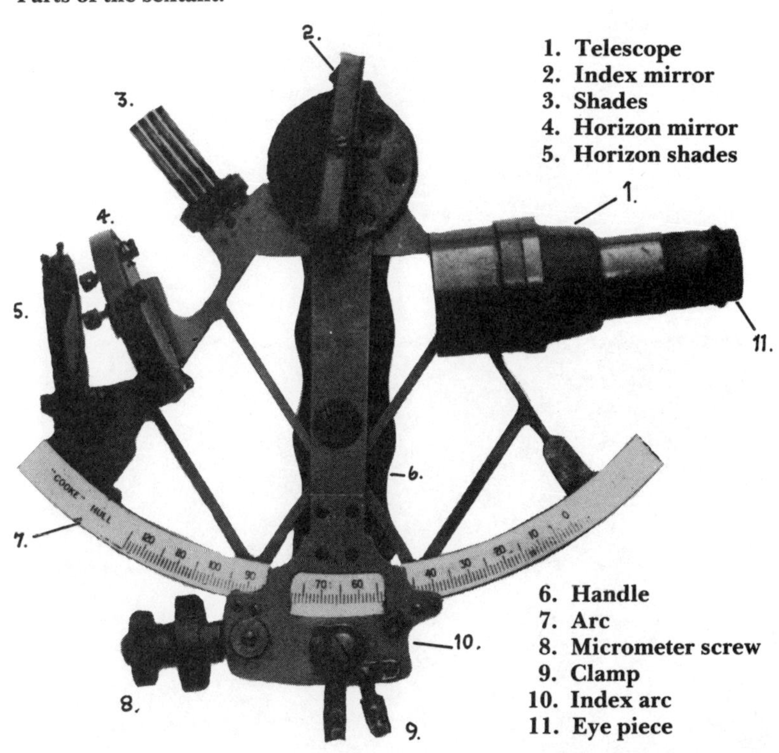

The instrument

The principal parts of the marine sextant are as follows:

1 *The frame*
The lightweight alloy structure on which the various parts are mounted.

2 *The arc*
The scale usually graduated in degrees.

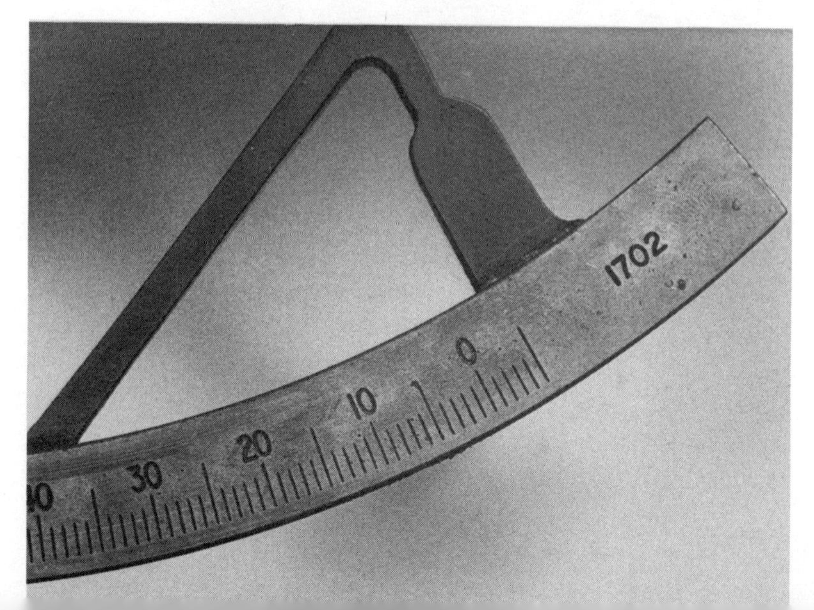

3 *The index arm*

This is the moveable arm of the instrument which is attached to one of the mirrors. As the mirror is adjusted to bring the sun to the horizon, the amount of this adjustment is recorded by the index arm moving along the graduated arc.

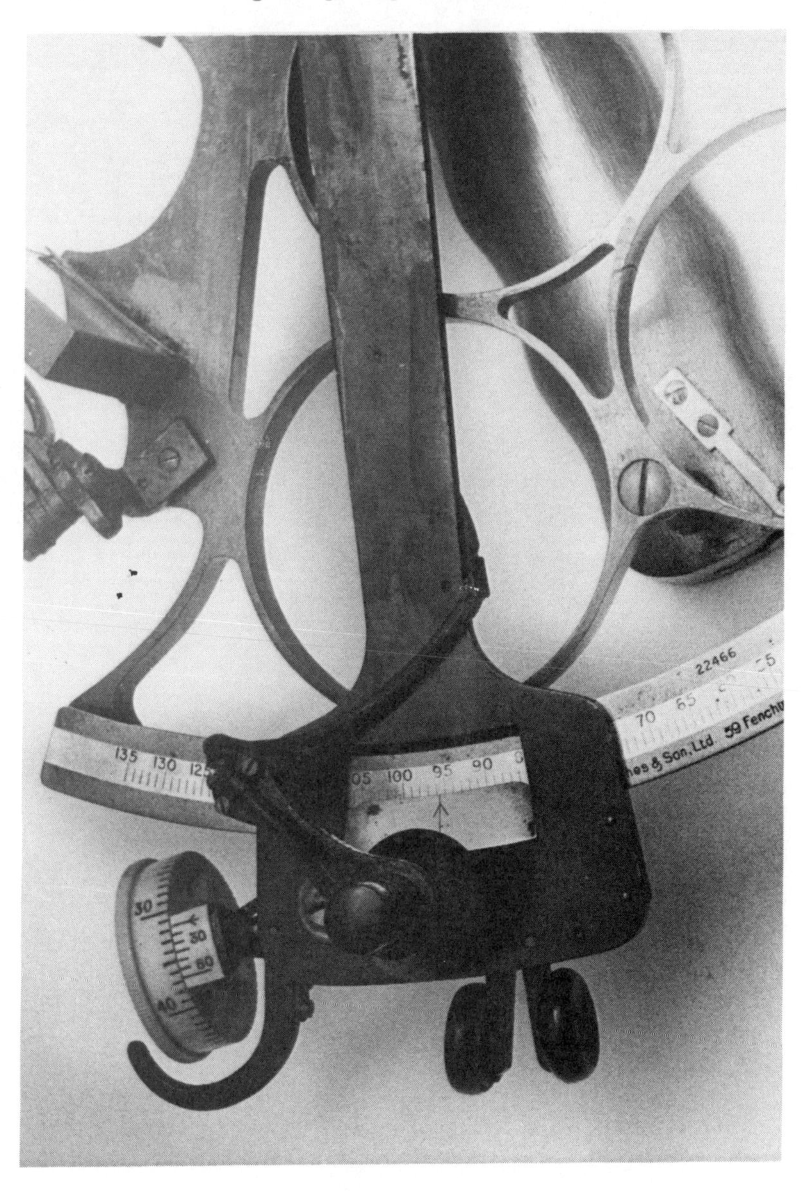

4 *The micrometer screw*
A fine adjustment drum, usually graduated in minutes to supplement the degrees measured on the arc.

5 *The telescope*
A normal telescope or 'monocular' which magnifies the celestial body being observed, thus making sight taking easier and more accurate.

6 *The index mirror*
This is the moving part of the instrument and reflects the heavenly body into the second mirror as the index arm, to which it is attached, is moved. As mentioned, the amount of this movement is recorded on the arc.

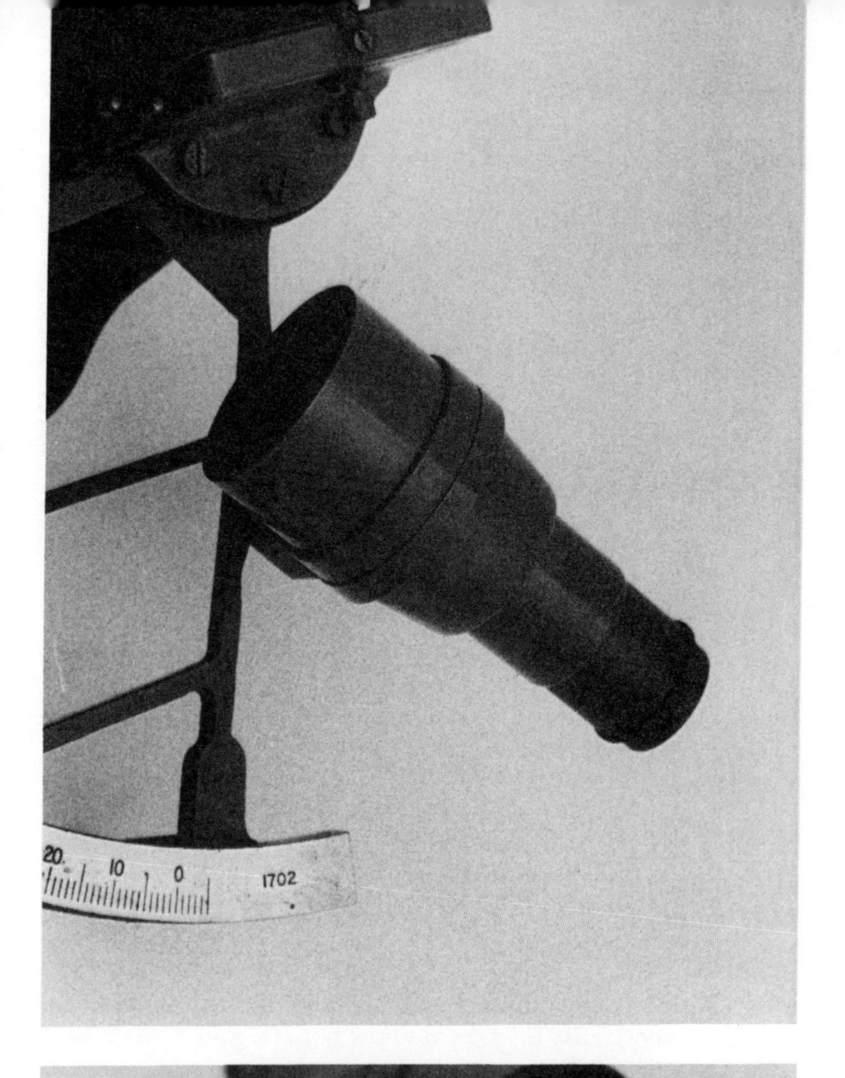
20
10
0
1702

7 *The horizon mirror*

This is actually only half a mirror, since the left-hand half, as seen through the telescope, is plain glass. The sun is reflected into the mirrored half by the index mirror and the horizon, when the sextant is pointed in that direction, is seen through the clear glass. The two are brought together in order to measure the altitude. Some recent sextants superimpose the sun on the horizon in the full mirror rather than splitting the mirror into two.

8 *Shades*

Colored or polaroid glass shades are inserted between the two mirrors to cut down the glare of the sun. The different intensity of shades allows for different intensity of glare. Similar, albeit lighter, shades are fitted behind the horizon mirror to reduce horizon glare.

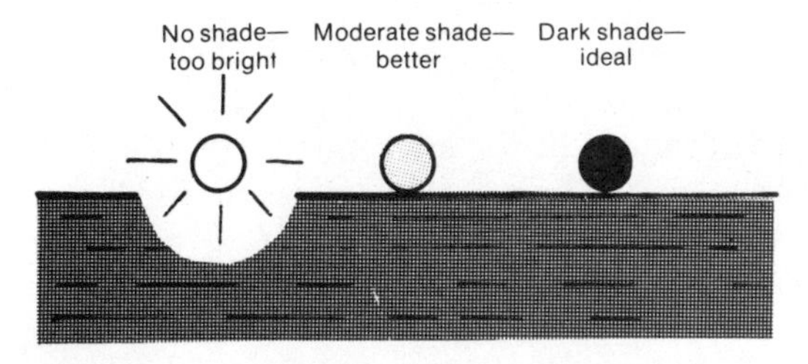

Using different shades sharpens the sun's image in the telescope and makes it easier to sight accurately on the horizon.

Holding the sextant

Most sextants are designed to be held in the right hand and to that effect a handle is attached to the right-hand side of the instrument. All fittings, mirrors, etc. are attached to the left-hand side. In use the instrument can become quite heavy despite its lightweight construction, particularly when attempting to 'shoot' three or

Holding the sextant correctly ensures good, accurate sights.

four stars. For this reason some support is necessary from the left hand and this is best done by placing the first finger under the forward end of the arc or the micrometer screw and the thumb under the rear end of the arc. This assists in the 'rocking' technique described later and also allows for easy adjustment of the micrometer screw.

Sighting through the telescope

When the heavenly body is in a position for taking a sight it will appear in the sextant telescope in the right-hand side of the horizon mirror. The horizon will appear in the left-hand side. When the sight is taken, as described later, the object and the horizon must be just touching. Thus, they must be brought together on the centerline of the horizon mirror, a procedure that involves moving the sextant until the heavenly body sits right on the centerline of the mirror and the horizon laps slightly over the centerline, allowing both to be brought together.

The correct location of sun and horizon in the horizon mirror prior to taking the sight.

Rocking the sextant

It is critically important that when a sight is taken, the sextant is exactly vertical to the horizon. This is not easy to determine by eye, so a 'spirit level' technique is used involving the sun (or other heavenly body) and the horizon. By supporting the sextant with the finger and thumb of the left hand as described earlier, the sextant is rocked out of the vertical first to one side and then to the other. The sun, seen through the telescope, appears to rise towards the top of the mirror as the sextant is tilted out of the vertical. By continuing this rocking action, the sun will be seen to describe an inverted arc rising each side of the mirror and reaching its lowest point when the sextant is exactly vertical.

The altitude reading is only accurate when taken with the sun at the bottom of this arc, and therefore the sextant must be rocked constantly while adjustments are made to bring the sun down to the horizon. This takes some practice, particularly with a star and even more particularly with a high altitude body, since the steepness of the arc decreases with higher altitudes.

Rocking the sextant out of the vertical enables it to be read with maximum accuracy.

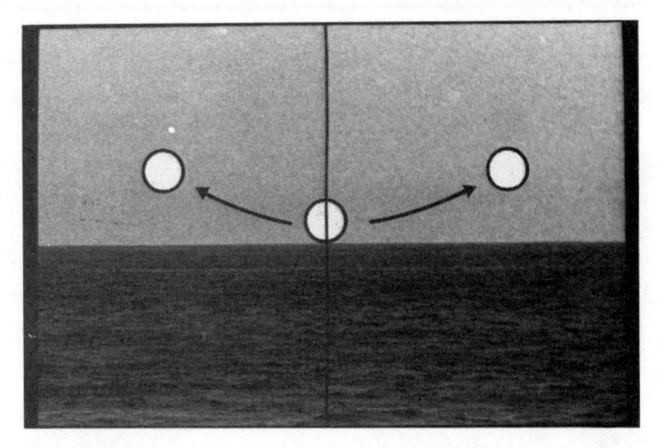

The reading is taken when the sun is at the bottom of the arc created by rocking the sextant.

Taking the sight

The horizon mirror is divided down the center, as described, with the horizon showing through clear glass on the left and the body reflected on the right. The sight must be taken with the body touching the horizon exactly on the centerline of the mirror, i.e. where the clear and mirrored glass join. This is done in conjunction with the rocking described above so that at the moment of sight taking, the heavenly body should be describing an arc, the bottom of which touches the horizon exactly on the centerline of the mirror. With the sun and moon, the *lower limb* (bottom edge) is taken for preference although this is not always possible with the moon, depending on its phase at the time. The *upper limb* can be taken, but it is more difficult to land accurately on the horizon and also needs special corrections. Stars and planets are too small to worry about upper or lower limb.

'Letting her rise'

Because of earth's rotation, heavenly bodies are constantly moving. When the sextant is pointed towards an easterly horizon the heavenly bodies are rising and moving upwards in the telescope as the sight is being taken. When facing west, they appear to be falling. This movement is quite pronounced and can complicate sight taking. Most experienced navigators prefer to adjust the instrument so that the heavenly body is just below the horizon (facing east) and allow the body to rise to the horizon, without

Cutting too high

Cutting too low

Right on!

Stars and planets (left) are too small to have an upper or lower limb. The sun (center) should always be taken with the lower limb, as should the moon unless this is not possible, when the upper limb must be used (right).

any further adjustment of the micrometer screw. In short, having established the heavenly body close to and below the horizon, no further adjustments are made, other than the rocking of the sextant. The reading is taken at the moment the body climbs to its exact location on the horizon. When taking sights towards a westerly horizon, of course, the body is poised *above* the horizon and allowed to fall without further adjustments to the micrometer screw.

'Bringing her down'

Beginners sometimes have difficulty in bringing a heavenly body down to the horizon. Particularly is this the case with stars and

Letting her rise. No adjustment should be made to the micrometer screw while the sun is allowed to rise to its own level on the horizon.

planets which are mere pinpoints in the sky. There are two methods employed to achieve this, the first requires a pre-computed altitude, the second is a simple practical system which is best described as follows:

1 Set the sextant at absolute zero on both micrometer screw and arc. Point the telescope of the sextant to the sun.

2 By moving the sextant slightly, two suns will be seen, one reflected and one true.

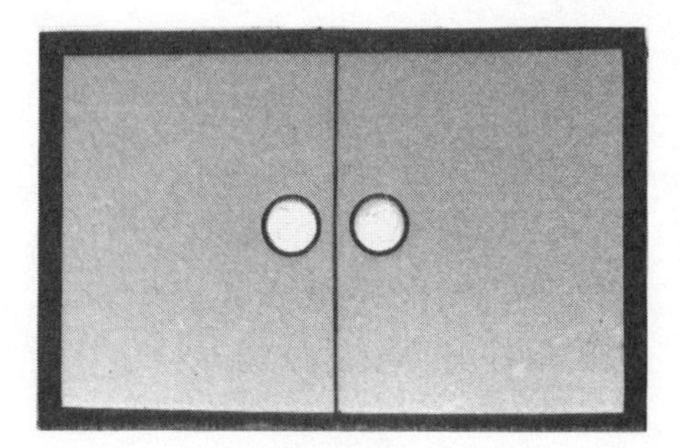

3 Lower the sextant slowly, adjusting the index arm so that the reflected sun is retained in the telescope. The true sun, of course, will disappear upwards as the sextant is lowered.

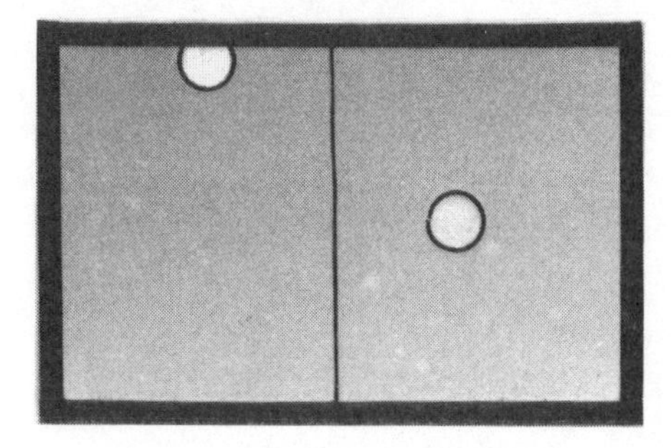

4 Continue lowering the sextant until the horizon appears in the left-hand side of the mirror at which point the index arm can be clamped into position. Final adjustment is made with the micrometer screw, rocking the sextant as described.

Take great care of the sextant, it is a precision instrument. It should always be stowed in its box.

Care of the sextant

Since it is a precision instrument, the sextant must be treated carefully and maintained periodically. Any severe bump or knock may upset the delicate balance of the mirrors. Similarly, excessive vibration or heat—such as when the sextant is left lying in the sun—will create problems. Treat the sextant like the expensive instrument it is and ensure that it is carefully stowed away when not in use. Special oil for lubrication will be provided with the instrument, and cleaning the mirrors and telescope must be done carefully and gently to prevent dislodging them from their precision-set locations.

Timing the sight

In most celestial computations, the timing of the sight is vitally important for accuracy. The position of each heavenly body in the sky is listed in *The Nautical Almanac* for each second of each minute of each day of the year. The position of the boat is found by relating it to the position of the body at the time the sight was taken.

An assistant taking the time of sight from a watch makes for greater accuracy than trying to do it all yourself.

Hence the importance of timing the sight to the nearest second. All time in navigation is *Greenwich Mean Time* (GMT) and one watch or chronometer on board should be kept on that time.

Until you gain experience the best system of timing the sight is to co-opt an assistant from the crew. Attempting to use the sextant and watch the chronometer at the same time will lead to confusion and possible error. One second out in the timing can considerably affect the result of the sight. Better to concentrate on the sextant and get the sight exactly right, and have an assistant read the time at the instant you take the sight. Later you may find it convenient to use a stop watch to time the interval between reading the sight and reading the chronometer. But this takes experience and initially two navigators are decidedly better than one.

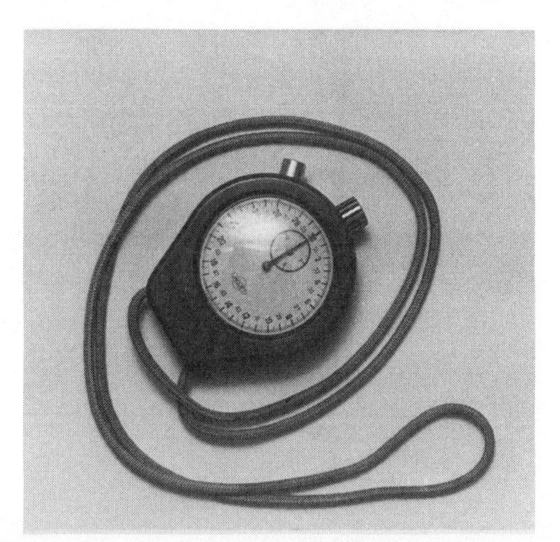

If you must take your own sight time, a stop watch is essential.

Sophisticated radio units are ideal for receiving time signals anywhere in the world.

Chronometers and watches

A chronometer is, in effect, an extremely accurate clock. Traditionally, navigation chronometers have been specially made time pieces mounted in wooden cases and protected against all possibility of damage, even that which may be due to fluctuations of temperature and humidity. However, with the advent of radio signals there is less need for such extreme accuracy in the navigator's clock, and for small boat work any reasonably accurate watch can be used, providing you have a radio on board that can be used to check the watch just prior to sight taking. Since accurate timing of sights is imperative the watch must be checked immediately before sight taking for even a one-second error will cause inaccuracies in the results of the sight.

Rating the watch

The argument against using an ordinary watch is that in the event of a radio failure or other problems preventing you from obtaining a radio check, the timing of your sights will not be sufficiently accurate. While this argument has a sound basis, modern radios and radio signals are virtually as reliable as a chronometer. However, as a precaution, a 'rate book' should be kept in which details of the gaining or losing characteristics of the watch are tabulated day by day. Then if a radio time check is not obtainable, the watch's gain or loss can be allowed so as to minimise the possibility of error.

In the event of loss of radio contact, the watch should be reliable enough that the rate of gain or loss recorded in the rate book will be consistent and the watch can be used with reasonable accuracy.

Radio time signals

Radio time signals are broadcast continuously from many stations throughout the world. D M A H T C Publications Nos 117A and 117B, *Radio Navigational Aids,* are volumes devoted entirely to details of radio signals and include particulars of time signals. A total list of all stations with their frequencies and the type of signal they broadcast are given in these volumes.

The station you tune into for time signals will depend greatly on your location. While many of the major stations can be picked up anywhere in the world, local stations often give a clearer signal by virtue of their closeness. The selection of the frequency for best reception will depend on your location, the time of day and atmospheric conditions. It is generally accepted that for long-range use the 15 MHz band is the best for daylight reception, and the 5 MHz band for night-time reception. Due to local atmospheric conditions radio signals may sometimes be blotted out or become difficult to interpret, in which case you should look for another station.

There are two forms of time signal, one using a voice and one a system of Morse or similar coded signals. The easiest to read are obviously the audio signals in which a time frequency is indicated by a 'pip' on each second of the minute, with a voice announcing the time during the last fifteen seconds of each minute. Station WWV broadcasts from Fort Collins, Colorado and station WWVH from Hawaii. Both broadcast across a range of frequencies from 2.5 to 20.0 MHz every hour of the day.

Thus the procedure for checking your navigating watch is to listen to the pips on each second comparing them with the second hand of your watch and waiting for the announcement of the time during the last fifteen seconds of each minute. This is no more difficult than listening to the time signal on the telephone except that your watch must be checked to the nearest second. For all other details related to time signals refer to the volumes mentioned above.

Chapter 3 Errors and Corrections

The altitude reading obtained from a sextant sight is subject to two types of errors:

Instrument errors
These are errors arising in the sextant itself due to misalignment of the mirrors or some other physical problem.

Sight errors
There are a number of errors inherent to sight taking due to factors such as the curvature of the earth, refraction of the sun's rays through the atmosphere, and so on. These must be taken into account before an accurate altitude can be obtained.

Instrument errors

Sextant errors fall into two categories; errors to the instrument which require it to be returned to an instrument maker or the factory, and 'working' errors which you can correct yourself on board the boat.

The accuracy of the sextant depends on the mirrors being exactly perpendicular to the plane of the instrument, and the faces of the mirrors being exactly parallel when the reading of the sextant is at zero. These mirrors are mounted in such a way that adjustments can be made to correct any misalignment caused by heat, vibration or physical damage. As a general rule they are mounted so that they can be adjusted by means of a screw on the back.

Since heat, humidity, vibration and, to a certain extent, physical damage, are an unavoidable part of using a sextant on board a boat, you should check the mirrors frequently to ensure that they are in alignment and adjust them if not. There are three basic errors which can creep into a sextant reading due to the mirror

Correct holding and handling of the sextant avoids physical errors.

problems and it is a wise practice, followed by most professional navigators, to run a quick check on these errors before taking any sights.

Index error

This is the most common error caused by mirror misalignment, and occurs if the faces of the two mirrors are not exactly parallel when the instrument is set at absolute zero. It can be easily measured and allowed for in sight calculations providing it is not too great. Indeed, if it is less than about 5′ of arc it should be allowed for in calculation rather than removed by adjusting the mirror. Adjustment of any of the mirrors tends to create further problems in that correcting one error can induce another. Better to allow for the error in the calculation unless it is sufficient to warrant a complete check and adjustment of both mirrors.

Finding the index error

This can be done in one of two ways, depending on whether the sextant is being checked by day or by night. Since the sextant is rarely used at night, the most common method is to check for index error during the daytime using the horizon. The procedure is as follows:

1 Set the sextant at absolute zero on the arc and micrometer screw. Look through the telescope directly at the horizon.

2 Both true and reflected horizons should appear in a straight line if no index error is present.

No index error

3 If the line of the horizon appears broken, adjust the micrometer screw until true and reflected horizons are in one straight line.

Index error present in instrument.

4 Read off the index error on the micrometer screw. This is the error to be applied to sights and is added if the reading is 'off the arc' (behind zero) and subtracted if the error is 'on the arc' (above zero).

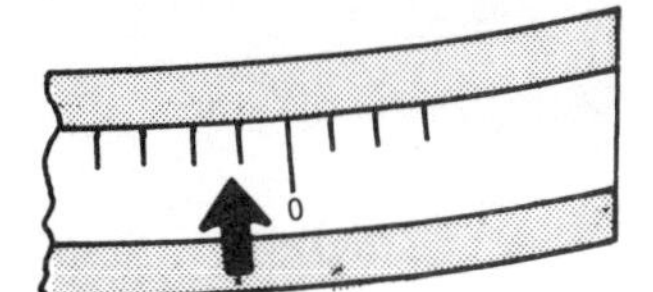

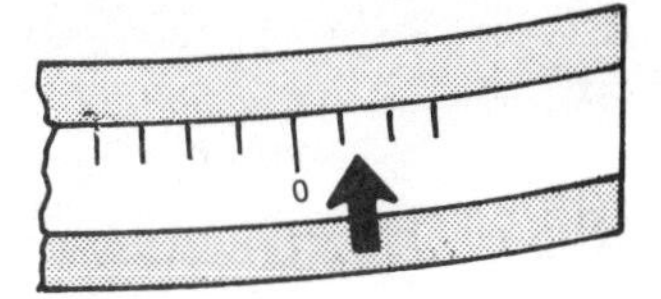

The index error is said to be 'on the arc' (left) or 'off the arc' (right).

5 If the reading is greater than 5′ it must be removed by adjusting the small screw closest to the body of the instrument on the back of the horizon mirror.

Side error

This error occurs when the horizon mirror is not absolutely perpendicular to the body of the instrument. As its name denotes, it can be found by tilting the sextant to one side. It cannot be measured, and therefore any significant error must be removed. The procedure for finding side error is as follows:

1 Set the sextant at absolute zero on both the arc and the micrometer screw.

2 Look through the telescope at the horizon, gradually leaning the sextant to one side until it is at an angle of about 45° to the vertical.

3 If the true and reflected horizons are not in one straight line, side error exists. Slight side error is acceptable, but any substantial error must be removed by adjusting the small screw farthest from the body of the instrument on the back of the horizon mirror.

No side error.

Side error is present and must be removed.

Error of perpendicularity

This error occurs when the index mirror is not perpendicular to the plane of the instrument. It cannot be measured and must be removed if it is significant. The procedure for finding the error of perpendicularity is somewhat more involved than the other two, and will need a certain amount of experimentation before the sextant is held correctly to make the error obvious. The procedure is as follows:

1 Set the index arm somewhere in the middle of the arc.
2 Turn the sextant round and look into the index mirror from above the instrument at an oblique angle.

Sighting up for the error of perpendicularity takes a bit of practice.

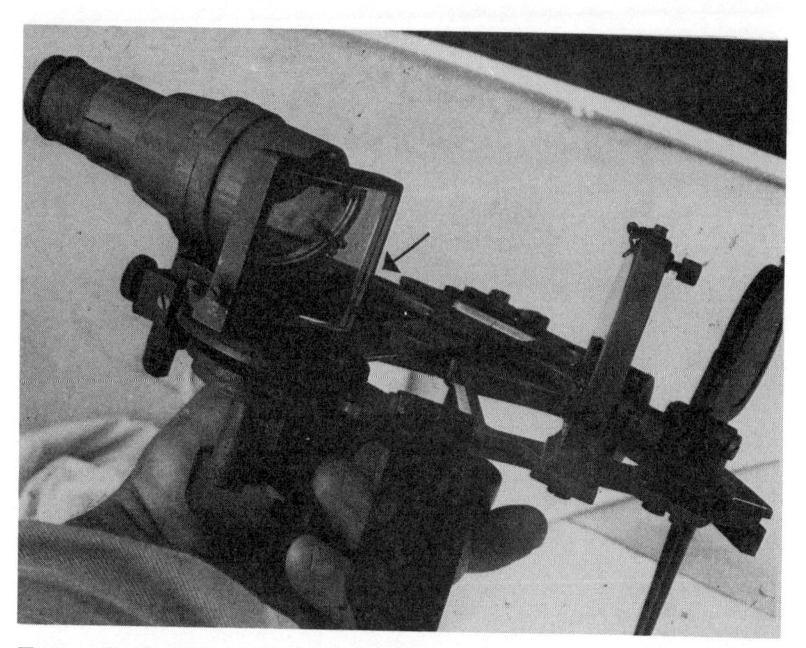

True and reflected arcs must coincide (arrow) if there is no perpendicularity.

3 If the reflected arc in the index mirror is in line with the true arc seen behind and beyond it, error of perpendicularity does not exist.

4 If the true and reflected arcs are out of alignment, the error must be adjusted by turning the small screw on the back of the index mirror.

Mirror adjustment is by means of small screws attached to the back of the mirrors.

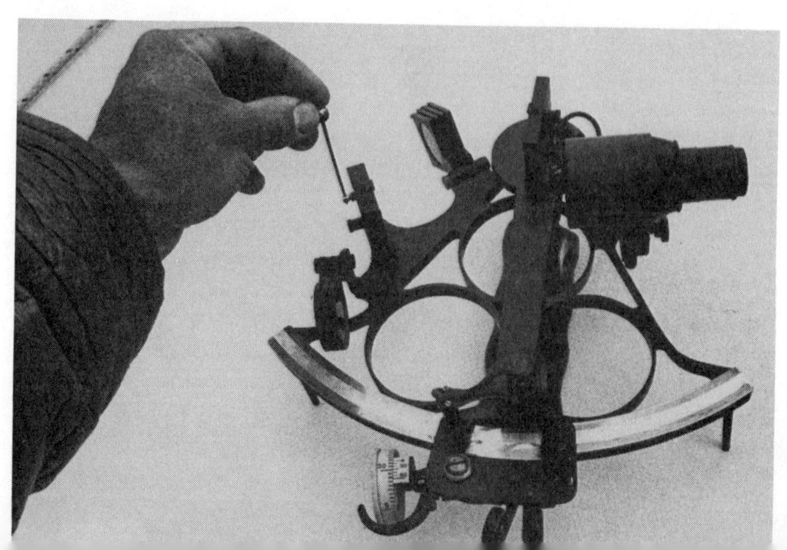

Adjustment of the mirrors is by trial and error, watching the horizon as each adjustment is carried out.

Mirror adjustments

It is most important when adjusting the mirrors for any of the three principal errors to ensure that one adjustment does not create another error. In short, since the errors are inter-related, when one mirror is adjusted, all errors must be checked. Indeed, it would be safe to say that rarely can one error be taken out without adjustments being required for the other two. These adjustments are all trial and error and can only be checked visually. For this reason, as mentioned, it is preferable to allow for index error as an addition or subtraction to a sight reading rather than attempt to adjust it all the time. The other errors, however, cannot be allowed for by calculation and must be removed by adjustment of the mirrors if they become significant.

Using a star

The first two errors can be found by using a star instead of the horizon. Index error is present when the sextant, set at absolute zero and pointed towards a star, indicates a true and reflected star separated one vertically above the other on the centerline of the horizon mirror. Side error exists when the true and reflected stars separate horizontally. Checking the sextant with a star is probably more accurate than with the horizon, but since these checks should be carried out routinely prior to taking a sight, the daytime method using the horizon is more common.

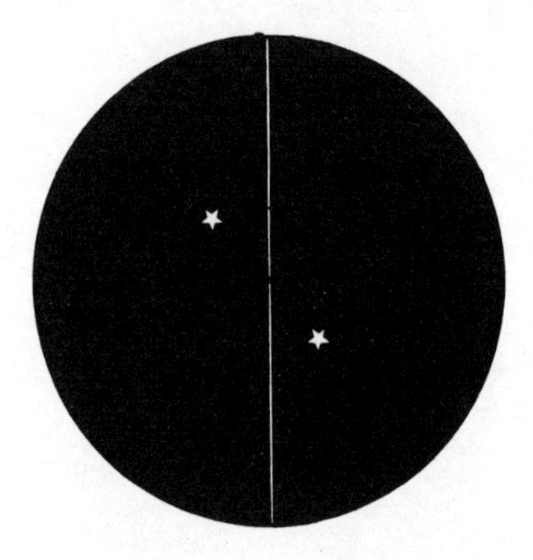

Using a star indicates two errors at once. Here both index and side errors exist.

Sight errors

There are a number of errors which occur when an altitude is obtained due to a variety of factors. I do not propose at this stage to go into detail on all these factors since they are numerous and many are included in one composite correction table. However, it is interesting to look as some of these errors and how they arise, together with the way in which they are corrected. The method described here is the one most commonly used in which two tables, both incorporated in *The Nautical Almanac*, are used.

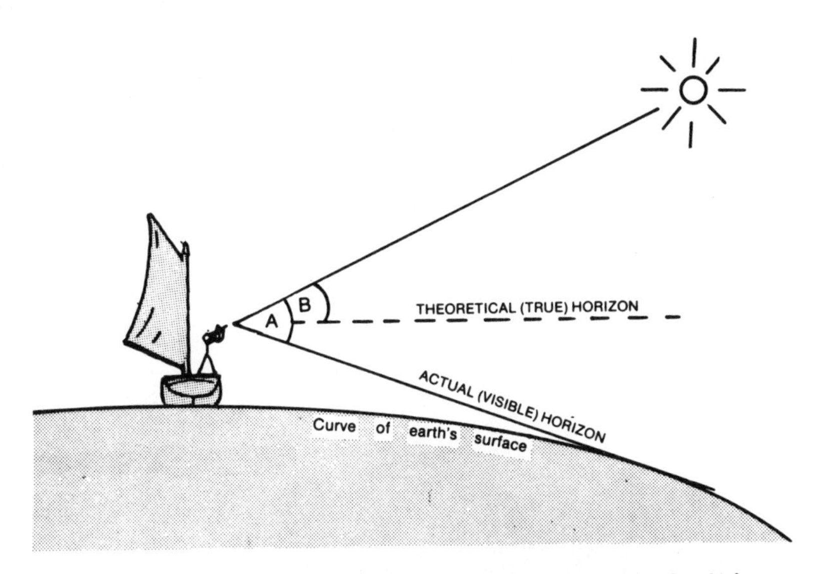

Because of the curve of the earth's surface, the sextant altitude (A) is incorrect. The true reading should be with the horizon at eye level (B). The difference is the dip.

Dip

More correctly termed *dip of the sea horizon*, this error comes about as a result of the curve of the earth's surface and the height above sea level at which the sight is taken. In theory, the sextant altitude measures the angle of the sun above the horizon. However, the higher you are above sea level the farther away the horizon appears. And the farther away the horizon, the more it curves over the earth's surface creating an error in the altitude reading.

Since the sextant altitude increases as the height above sea level increases, the error of dip is always subtracted from the altitude reading. A table on the inside cover of *The Nautical Almanac* headed DIP gives the amount of error to be subtracted from the sextant reading according to the height-of-eye at which the sight was taken. Height-of-eye can be roughly gauged by the height of the deck above the water plus your height as the navigator taking the sight.

Altitude correction

This is the heading of the table which offers a single correction for all the remaining sight errors. Most of these errors are to do with the location of the heavenly body, its size and the atmosphere

DIP

Ht. of Eye	Corrⁿ	Ht. of Eye
m	′	ft.
2·4		8·0
	−2·8	
2·6		8·6
	−2·9	
2·8		9·2
	−3·0	
3·0		9·8
	−3·1	
3·2		10·5
	−3·2	
3·4		11·2
	−3·3	
3·6		11·9
	−3·4	
3·8		12·6
	−3·5	
4·0		13·3
	−3·6	
4·3		14·1
	−3·7	
4·5		14·9
	−3·8	
4·7		15·7
	−3·9	
5·0		16·5
	−4·0	
5·2		17·4
	−4·1	
5·5		18·3
	−4·2	
5·8		19·1
	−4·3	
6·1		20·1
	−4·4	
6·3		21·0
	−4·5	
6·6		22·0
	−4·6	
6·9		22·9
	−4·7	
7·2		23·9
	−4·8	
7·5		24·9
	−4·9	
7·9		26·0
	−5·0	
8·2		27·1
	−5·1	
8·5		28·1
	−5·2	
8·8		29·2
	−5·3	
9·2		30·4
	−5·4	
9·5		31·5
	−5·5	
9·9		32·7
	−5·6	
10·3		33·9
	−5·7	
10·6		35·1
	−5·8	
11·0		36·3
	−5·9	
11·4		37·6
	−6·0	
11·8		38·9
	−6·1	
12·2		40·1
	−6·2	
12·6		41·5
	−6·3	
13·0		42·8
	−6·4	
13·4		44·2
	−6·5	
13·8		45·5
	−6·6	
14·2		46·9
	−6·7	
14·7		48·4
	−6·8	
15·1		49·8
	−6·9	
15·5		51·3
	−7·0	
16·0		52·8
	−7·1	
16·5		54·3
	−7·2	
16·9		55·8
	−7·3	
17·4		57·4
	−7·4	
17·9		58·9
	−7·5	
18·4		60·5
	−7·6	
18·8		62·1
	−7·7	
19·3		63·8
	−7·8	
19·8		65·4
	−7·9	
20·4		67·1
	−8·0	
20·9		68·8
	−8·1	
21·4		70·5

Ht. of Eye	Corrⁿ
m	′
1·0	−1·8
1·5	−2·2
2·0	−2·5
2·5	−2·8
3·0	−3·0
See table ←	
m	′
20	−7·9
22	−8·3
24	−8·6
26	−9·0
28	−9·3
30	−9·6
32	−10·0
34	−10·3
36	−10·6
38	−10·8
40	−11·1
42	−11·4
44	−11·7
46	−11·9
48	−12·2
ft.	′
2	−1·4
4	−1·9
6	−2·4
8	−2·7
10	−3·1
See table ←	
ft.	′
70	−8·1
75	−8·4
80	−8·7
85	−8·9
90	−9·2
95	−9·5
100	−9·7
105	−9·9
110	−10·2
115	−10·4
120	−10·6
125	−10·8
130	−11·1
135	−11·3
140	−11·5
145	−11·7
150	−11·9
155	−12·1

Dip tables from *The Nautical Almanac*.

Refraction of the sun by atmospheric layers makes it appear other than where it is and gives false sextant reading.

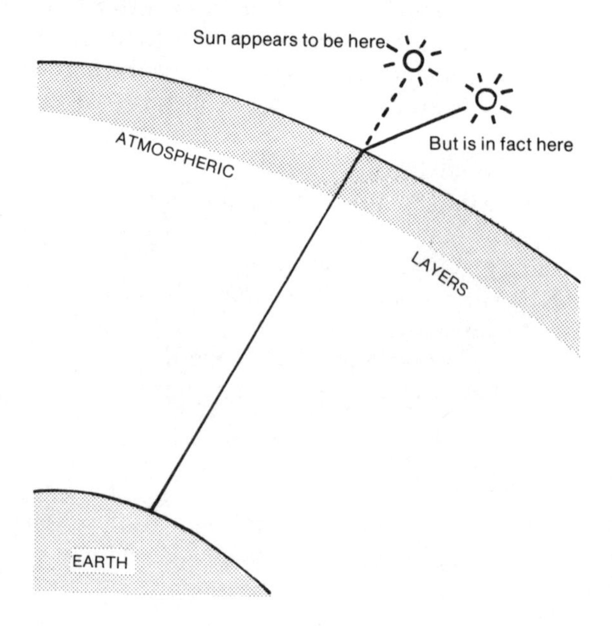

around the earth. *Refraction*, for example, is the result of the bending of the sun's rays as they penetrate earth's atmosphere from outer space, rather like the apparent bending of a stick when placed in a bucket of water. *Parallax* is the error resulting from the fact that the sight was taken from the earth's surface whereas all calculations are made from the earth's center. *Semidiameter*, an error which applies to the sun in particular, comes about because the lower limb of the sun is used for convenience in bringing it down to the horizon when taking the sight. All calculations relating to the sun, however, are made from its center, and therefore the difference between the lower edge and the center (the semidiameter) must be allowed as a correction. The Altitude Correction table covers other errors but these are of relative insignificance.

ALTITUDE CORRECTION TABLES 10°-90°

SUN

OCT.—MAR. App. Alt.	Lower Limb	Upper Limb
9 34	+10·8	−21·5
9 45	+10·9	−21·4
9 56	+11·0	−21·3
10 08	+11·1	−21·2
10 21	+11·2	−21·1
10 34	+11·3	−21·0
10 47	+11·4	−20·9
11 01	+11·5	−20·8
11 15	+11·6	−20·7
11 30	+11·7	−20·6
11 46	+11·8	−20·5
12 02	+11·9	−20·4
12 19	+12·0	−20·3
12 37	+12·1	−20·2
12 55	+12·2	−20·1
13 14	+12·3	−20·0
13 35	+12·4	−19·9
13 56	+12·5	−19·8
14 18	+12·6	−19·7
14 42	+12·7	−19·6
15 06	+12·8	−19·5
15 32	+12·9	−19·4
15 59	+13·0	−19·3
16 28	+13·1	−19·2
16 59	+13·2	−19·1
17 32	+13·3	−19·0
18 06	+13·4	−18·9
18 42	+13·5	−18·8
19 21	+13·6	−18·7
20 03	+13·7	−18·6
20 48	+13·8	−18·5
21 35	+13·9	−18·4
22 26	+14·0	−18·3
23 22	+14·1	−18·2
24 21	+14·2	−18·1
25 26	+14·3	−18·0
26 36	+14·4	−17·9
27 52	+14·5	−17·8
29 15	+14·6	−17·7
30 46	+14·7	−17·6
32 26	+14·8	−17·5
34 17	+14·9	−17·4
36 20	+15·0	−17·3
38 36	+15·1	−17·2
41 08	+15·2	−17·1
43 59	+15·3	−17·0
47 10	+15·4	−16·9
50 46	+15·5	−16·8
54 49	+15·6	−16·7
59 23	+15·7	−16·6
64 30	+15·8	−16·5
70 12	+15·9	−16·4
76 26	+16·0	−16·3
83 05	+16·1	−16·2
90 00		

APR.—SEPT. App. Alt.	Lower Limb	Upper Limb
9 39	+10·6	−21·2
9 51	+10·7	−21·1
10 03	+10·8	−21·0
10 15	+10·9	−20·9
10 27	+11·0	−20·8
10 40	+11·1	−20·7
10 54	+11·2	−20·6
11 08	+11·3	−20·5
11 23	+11·4	−20·4
11 38	+11·5	−20·3
11 54	+11·6	−20·2
12 10	+11·7	−20·1
12 28	+11·8	−20·0
12 46	+11·9	−19·9
13 05	+12·0	−19·8
13 24	+12·1	−19·7
13 45	+12·2	−19·6
14 07	+12·3	−19·5
14 30	+12·4	−19·4
14 54	+12·5	−19·3
15 19	+12·6	−19·2
15 46	+12·7	−19·1
16 14	+12·8	−19·0
16 44	+12·9	−18·9
17 15	+13·0	−18·8
17 48	+13·1	−18·7
18 24	+13·2	−18·6
19 01	+13·3	−18·5
19 42	+13·4	−18·4
20 25	+13·5	−18·3
21 11	+13·6	−18·2
22 00	+13·7	−18·1
22 54	+13·8	−18·0
23 51	+13·9	−17·9
24 53	+14·0	−17·8
26 00	+14·1	−17·7
27 13	+14·2	−17·6
28 33	+14·3	−17·5
30 00	+14·4	−17·4
31 35	+14·5	−17·3
33 20	+14·6	−17·2
35 17	+14·7	−17·1
37 26	+14·8	−17·0
39 50	+14·9	−16·9
42 31	+15·0	−16·8
45 31	+15·1	−16·7
48 55	+15·2	−16·6
52 44	+15·3	−16·5
57 02	+15·4	−16·4
61 51	+15·5	−16·3
67 17	+15·6	−16·2
73 16	+15·7	−16·1
79 43	+15·8	−16·0
86 32	+15·9	−15·9
90 00		

STARS AND PLANETS

App. Alt.	Corrⁿ
9 56	−5·3
10 08	−5·2
10 20	−5·1
10 33	−5·0
10 46	−4·9
11 00	−4·8
11 14	−4·7
11 29	−4·6
11 45	−4·5
12 01	−4·4
12 18	−4·3
12 35	−4·2
12 54	−4·1
13 13	−4·0
13 33	−3·9
13 54	−3·8
14 16	−3·7
14 40	−3·6
15 04	−3·5
15 30	−3·4
15 57	−3·3
16 26	−3·2
16 56	−3·1
17 28	−3·0
18 02	−2·9
18 38	−2·8
19 17	−2·7
19 58	−2·6
20 42	−2·5
21 28	−2·4
22 19	−2·3
23 13	−2·2
24 11	−2·1
25 14	−2·0
26 22	−1·9
27 36	−1·8
28 56	−1·7
30 24	−1·6
32 00	−1·5
33 45	−1·4
35 40	−1·3
37 48	−1·2
40 08	−1·1
42 44	−1·0
45 36	−0·9
48 47	−0·8
52 18	−0·7
56 11	−0·6
60 28	−0·5
65 08	−0·4
70 11	−0·3
75 34	−0·2
81 13	−0·1
87 03	0·0
90 00	

App. Alt.	Additional Corrⁿ
1984 VENUS	
Jan. 1-Dec. 12	
0	
	+0·1
60	
Dec. 13-Dec. 31	
0	
	+0·2
41	
	+0·1
76	
MARS	
Jan. 1-Mar. 4	
0	
	+0·1
60	
Mar. 5-Apr. 24	
0	
	+0·2
41	
	+0·1
76	
Apr. 25-June 15	
0	
	+0·3
34	
	+0·2
60	
	+0·1
80	
June 16-Aug. 27	
0	
	+0·2
41	
	+0·1
76	
Aug. 28-Dec. 31	
0	
	+0·1
60	

Altitude Correction Tables from *The Nautical Almanac*.

The correction tables for the sun are divided into two, one for the period October-March and the other for the period April-September. Stars do not have a monthly correction, but planets and the moon have corrections related to their movements at certain times of the year and these are indicated in the table. The Altitude Correction table is entered with the Apparent Altitude which, as will be seen later, is the sextant altitude corrected for all other errors. The correction is added in the case of sun and planets, and subtracted in the case of stars.

Correcting the sextant reading

Although professional navigators may adopt their own methods of correcting the sextant altitude, the most common method in practice follows the pattern:

1 Correction for index error (if any)
2 Correction for dip
3 Correction for all other errors (Altitude Correction tables)

The following form indicates this method of correcting the sextant altitude to obtain the true altitude, and it will be seen that terminology in relation to the altitude is along the following lines:

Form for correcting the sextant altitude.

Sextant

Sext alt		
I.E.	________	±
Obs alt		
Dip	________	−
App Alt		
Alt Corr	________	±
TRUE ALT	________	(Ho)

Sextant altitude The altitude reading from the sextant.
Observed altitude The sextant altitude corrected for sextant errors.
Apparent altitude The sextant altitude corrected for sextant errors and for dip.
True altitude The sextant altitude corrected for all errors (code named Ho).

Since only the True Altitude can be used for sight computations it follows that *every* sextant reading must be corrected in this way before being used.

Chapter 4 Estimated Position and Dead Reckoning

The term *dead reckoning*, usually abbreviated to DR, is derived from *deduced reckoning*, the system by which the position of a boat is deduced or calculated from information on her past movements rather than by visual sights or observations. The DR position, always stated in latitude and longitude, is also sometimes called the *estimated position* (EP) is which case additional allowances are made for the effect of wind and currents.

There are two methods of obtaining an estimated position.

1 By plotting the boat's courses and distances (with allowances for all factors such as current, wind, etc which may have affected her) on the chart.

2 By calculating trigonometrically her change in latitude and longitude from a previous fix position, allowing for the same factors mentioned above.

EP by plotting

This is the simplest method of arriving at a boat's EP position, since it involves simply plotting the estimated courses and distances she has sailed with allowance for influencing factors from her last known fix position. It is the most common system used in coastal navigation but becomes less accurate when using small scale charts of the oceans since pegging out the distance run along a courseline can be difficult using the small latitude scale as mileage. A run of, say, twenty miles will hardly be measurable on the latitude scale of a small scale chart. Since an EP is required before each sight is taken, short runs are not uncommon, and yet must be done accurately if the best possible EP is to be obtained—a necessary factor in ensuring accurate results from sights. A larger scale chart can be used and the EP plotted on this—as in the case when plotting stars—but experienced navigators find

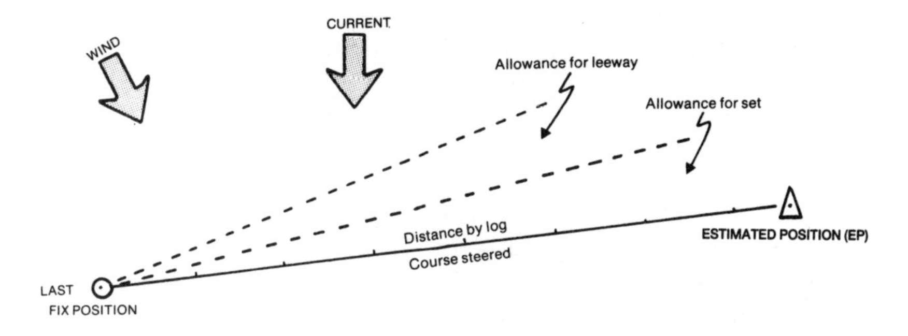

The EP is the most accurate position available working from the last known fix and using all known factors affecting the boat's progress.

calculation of the EP much quicker and more accurate.

If a plotting sheet is used, the last known fix is marked on the sheet according to latitude and longitude and the course the boat has since followed is laid off from it. By measuring along this course the log distance run from the fix and allowing for any known factors such as set of current or leeway, the EP can be plotted and then the co-ordinates of latitude and longitude taken off.

DR or EP by calculation

Although most people tend to be frightened of mathematical calculations, the method of obtaining a DR or EP by use of tables is simple and involves no more than the addition or subtraction of two lines of figures. As with all navigational problems, it is more a question of getting used to using tables rather than the actual mathematical calculation.

This applies to DR or EP by calculation, which can be done by a professional navigator in seconds, since he is used to the tables. The beginner, however, will find a few problems adjusting to these tables since, like so many navigational tables, they are compressed to make them compact and this tends to make them a little confusing. There are a number of different ways of calculating a DR position, all involving the use of a simple right-angled triangle. The commonly used table, which will be found in all volumes of nautical tables, is the Traverse Table. In this book the tables referred to are those of the *American Practical Navigator* (Bowditch).

The Traverse Table

This table is actually three tables combined and an explanatory 'box' at the foot of the page indicates the headings for each table. Headings for the three columns used in the first table are Dist (Distance), DLat (Difference of Latitude) and Dep (Departure). The second table uses only two columns headed DLo (Difference of Longitude) and Dep (Departure). These headings refer to the data required for calculating the DR position. A second 'box' at the top and bottom of each page indicates the boat's course in each quadrant of the compass.

The major heading in degrees at the top and bottom of the page serves two tables: In the first it is the cardinal heading of the boat and in the second table it represents the mean or middle latitude. The third table is not used for DR or EP calculations.

325° | 035° / 215° | 145° — TABLE 3 — Traverse **35°** Table — 325° | 035° / 215° | 145°

Dist	D Lat	Dep	Dist	D Lat	Dep	Dist	D Lat	Dep	Dist	D Lat	Dep	Dist	D Lat	Dep
1	0.8	0.6	61	50.0	35.0	121	99.1	69.4	181	148.3	103.8	241	197.4	138.2
2	1.6	1.1	62	50.8	35.6	22	99.9	70.0	82	149.1	104.4	42	198.2	138.8
3	2.5	1.7	63	51.6	36.1	23	100.8	70.5	83	149.9	105.0	43	199.1	139.4
4	3.3	2.3	64	52.4	36.7	24	101.6	71.1	84	150.7	105.5	44	199.9	140.0
5	4.1	2.9	65	53.2	37.3	25	102.4	71.7	85	151.5	106.1	45	200.7	140.5
6	4.9	3.4	66	54.1	37.9	26	103.2	72.3	86	152.4	106.7	46	201.5	141.1
7	5.7	4.0	67	54.9	38.4	27	104.0	72.8	87	153.2	107.3	47	202.3	141.7
8	6.6	4.6	68	55.7	39.0	28	104.9	73.4	88	154.0	107.8	48	203.1	142.2
9	7.4	5.2	69	56.5	39.6	29	105.7	74.0	89	154.8	108.4	49	204.0	142.8
10	8.2	5.7	70	57.3	40.2	30	106.5	74.6	90	155.6	109.0	50	204.8	143.4
11	9.0	6.3	71	58.2	40.7	131	107.3	75.1	191	156.5	109.6	251	205.6	144.0
12	9.8	6.9	72	59.0	41.3	32	108.1	75.7	92	157.3	110.1	52	206.4	144.5
13	10.6	7.5	73	59.8	41.9	33	108.9	76.3	93	158.1	110.7	53	207.2	145.1
14	11.5	8.0	74	60.6	42.4	34	109.8	76.9	94	158.9	111.3	54	208.1	145.7
15	12.3	8.6	75	61.4	43.0	35	110.6	77.4	95	159.7	111.8	55	208.9	146.3
16	13.1	9.2	76	62.3	43.6	36	111.4	78.0	96	160.6	112.4	56	209.7	146.8
17	13.9	9.8	77	63.1	44.2	37	112.2	78.6	97	161.4	113.0	57	210.5	147.4
18	14.7	10.3	78	63.9	44.7	38	113.0	79.2	98	162.2	113.6	58	211.3	148.0
19	15.6	10.9	79	64.7	45.3	39	113.9	79.7	99	163.0	114.1	59	212.2	148.6
20	16.4	11.5	80	65.5	45.9	40	114.7	80.3	200	163.8	114.7	60	213.0	149.1
21	17.2	12.0	81	66.4	46.5	141	115.5	80.9	201	164.6	115.3	261	213.8	149.7
22	18.0	12.6	82	67.2	47.0	42	116.3	81.4	02	165.5	115.9	62	214.6	150.3
23	18.8	13.2	83	68.0	47.6	43	117.1	82.0	03	166.3	116.4	63	215.4	150.9
24	19.7	13.8	84	68.8	48.2	44	118.0	82.6	04	167.1	117.0	64	216.3	151.4
25	20.5	14.3	85	69.6	48.8	45	118.8	83.2	05	167.9	117.6	65	217.1	152.0
26	21.3	14.9	86	70.4	49.3	46	119.6	83.7	06	168.7	118.2	66	217.9	152.6
27	22.1	15.5	87	71.3	49.9	47	120.4	84.3	07	169.6	118.7	67	218.7	153.1
28	22.9	16.1	88	72.1	50.5	48	121.2	84.9	08	170.4	119.3	68	219.5	153.7
29	23.8	16.6	89	72.9	51.0	49	122.1	85.5	09	171.2	119.9	69	220.4	154.3
30	24.6	17.2	90	73.7	51.6	50	122.9	86.0	10	172.0	120.5	70	221.2	154.9
31	25.4	17.8	91	74.5	52.2	151	123.7	86.6	211	172.8	121.0	271	222.0	155.4
32	26.2	18.4	92	75.4	52.8	52	124.5	87.2	12	173.7	121.6	72	222.8	156.0
33	27.0	18.9	93	76.2	53.3	53	125.3	87.8	13	174.5	122.2	73	223.6	156.6
34	27.9	19.5	94	77.0	53.9	54	126.1	88.3	14	175.3	122.7	74	224.4	157.2
35	28.7	20.1	95	77.8	54.5	55	127.0	88.9	15	176.1	123.3	75	225.3	157.7
36	29.5	20.6	96	78.6	55.1	56	127.8	89.5	16	176.9	123.9	76	226.1	158.3
37	30.3	21.2	97	79.5	55.6	57	128.6	90.1	17	177.8	124.5	77	226.9	158.9
38	31.1	21.8	98	80.3	56.2	58	129.4	90.6	18	178.6	125.0	78	227.7	159.5
39	31.9	22.4	99	81.1	56.8	59	130.2	91 2	19	179.4	125.6	79	228.5	160.0
40	32.8	22.9	100	81.9	57.4	60	131.1	91.8	20	180.2	126.2	80	229.4	160.6
41	33.6	23.5	101	82.7	57.9	161	131.9	92.3	221	181.0	126.8	281	230.2	161.2
42	34.4	24.1	02	83.6	58.5	62	132.7	92.9	22	181.9	127.3	82	231.0	161.7
43	35.2	24.7	03	84.4	59.1	63	133.5	93.5	23	182.7	127.9	83	231.8	162.3
44	36.0	25.2	04	85.2	59.7	64	134.3	94.1	24	183.5	128.5	84	232.6	162.9
45	36.9	25.8	05	86.0	60.2	65	135.2	94.6	25	184.3	129.1	85	233.5	163.5
46	37.7	26.4	06	86.8	60.8	66	136.0	95.2	26	185.1	129.6	86	234.3	164.0
47	38.5	27.0	07	87.6	61.4	67	136.8	95.8	27	185.9	130.2	87	235.1	164.6
48	39.3	27.5	08	88.5	61.9	68	137.6	96.4	28	186.8	130.8	88	235.9	165.2
49	40.1	28.1	09	89.3	62.5	69	138.4	96.9	29	187.6	131.3	89	236.7	165.8
50	41.0	28.7	10	90.1	63.1	70	139.3	97.5	30	188.4	131.9	90	237.6	166.3
51	41.8	29.3	111	90.9	63.7	171	140.1	98.1	231	189.2	132.5	291	238.4	166.9
52	42.6	29.8	12	91.7	64.2	72	140.9	98.7	32	190.0	133.1	92	239.2	167.5
53	43.4	30.4	13	92.6	64.8	73	141.7	99.2	33	190.9	133.6	93	240.0	168.1
54	44.2	31.0	14	93.4	65.4	74	142.5	99.8	34	191.7	134.2	94	240.8	168.6
55	45.1	31.5	15	94.2	66.0	75	143.4	100.4	35	192.5	134.8	95	241.6	169.2
56	45.9	32.1	16	95.0	66.5	76	144.2	100.9	36	193.3	135.4	96	242.5	169.8
57	46.7	32.7	17	95.8	67.1	77	145.0	101.5	37	194.1	135.9	97	243.3	170.4
58	47.5	33.3	18	96.7	67.7	78	145.8	102.1	38	195.0	136.5	98	244.1	170.9
59	48.3	33.8	19	97.5	68.3	79	146.6	102.7	39	195.8	137.1	99	244.9	171.5
60	49.1	34.4	20	98.3	68.8	80	147.4	103.2	40	196.6	137.7	300	245.7	172.1
Dist	Dep	D Lat	Dist	Dep	D Lat	Dist	Dep	D Lat	Dist	Dep	D Lat	Dist	Dep	D Lat

305° | 055° / 235° | 125° — **55°**

Dist.	D. Lat.	Dep.
D Lo	Dep.	
	m	D Lo

Excerpt from Traverse Tables.

The data

DLat

As its name denotes, DLat is the difference between two latitudes expressed in minutes. Thus the DLat between Lat 10°20′S and Lat 11°10′S is 50′S. The DLat between 10°20′S and 9°30′S is 50′N.

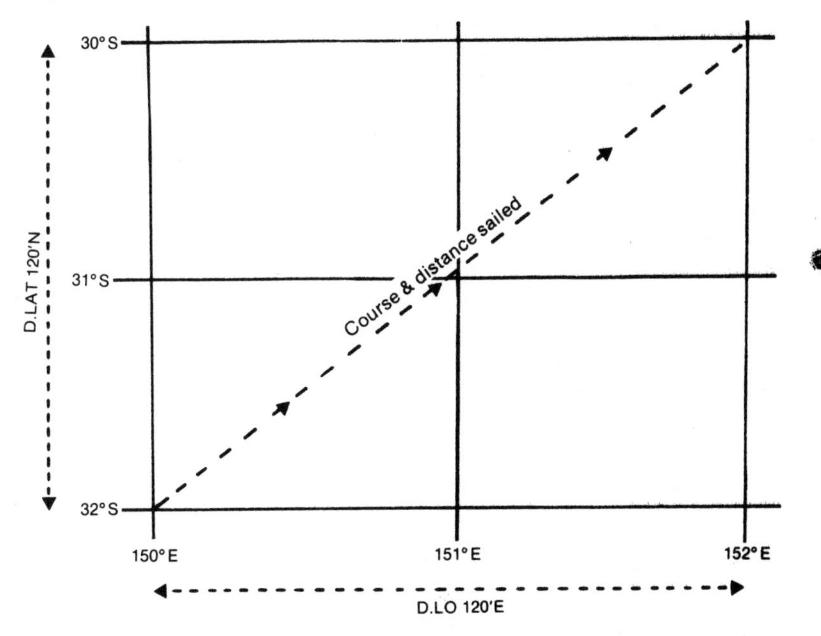

While sailing along a certain course for a certain distance, the boat creates DLat and DLo.

DLo

Similarly, the DLo is the difference in minutes of longitude between one meridian of longitude and another, measured E or W.

Dep

Departure is somewhat of a mystery factor in the DR calculation, but can generally be considered as minutes of DLo expressed in miles. However, that may be confusing to the beginner and since Dep is used only as a stepping stone from one table to another and it is not used in the actual DR workings, it can be ignored in terms of definition.

The method

Assume that you have established the boat's position some time before and wish now to run up a DR or EP position in order to establish her position prior to taking another sight. This means, in effect, converting the course and distance she has sailed into minutes of DLat and DLo. By then applying these to the original position latitude and longitude, the DR position latitude and longitude are obtained. The procedure for this is as follows:

Our boat was in a position Lat 35°20′S Long 151°25′E a few hours ago. Since then we have sailed a course of 035°T for a log distance of 44 miles. What is our DR lat and long position?

1 Write down the latitude and longitude of the original fix position (Lat 35°20′S Long 151°25′E).

2 Enter the Traverse Table with the course steered from this position. It will be found at the top or bottom of a page (035°).

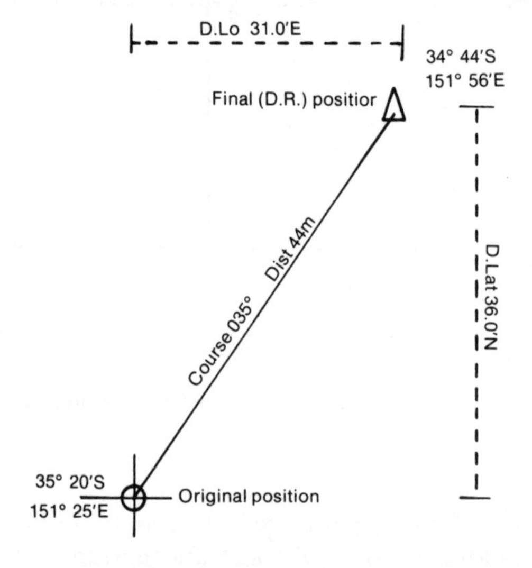

3 On this page, using the columns of the first table, enter the Dist column and find the distance run by log from the original fix position (44 miles).
Note: If the course is found at the bottom of the page all columns are read from the bottom.

4 Against this distance read out the adjacent figures for DLat (36.0′N) and Dep (25.2′).

5 Apply the DLat to the original fix lat to obtain the DR Lat (like names add, unlike names subtract)—(34°44′S).

6 Take the mean latitude (half way between original and DR Lats, to the nearest degree) and find the page on which it appears (35°S).

7 With the Dep just found (25.2′) and using the column headings for the second table, enter the Dep column. Against it in the DLo column read out the DLo (31.0′E to the nearest minute).

8 Apply this DLo to the original fix longitude to obtain the DR longitude (151°56′E).

Note that because the Traverse Table are condensed, the columns can be read from top or bottom of the page, according to where the course or mean latitude is found. To find the EP, the set and drift of the current is applied as an extra course and distance, the DLat and DLo being taken from the tables as described and then applied to the DR position.

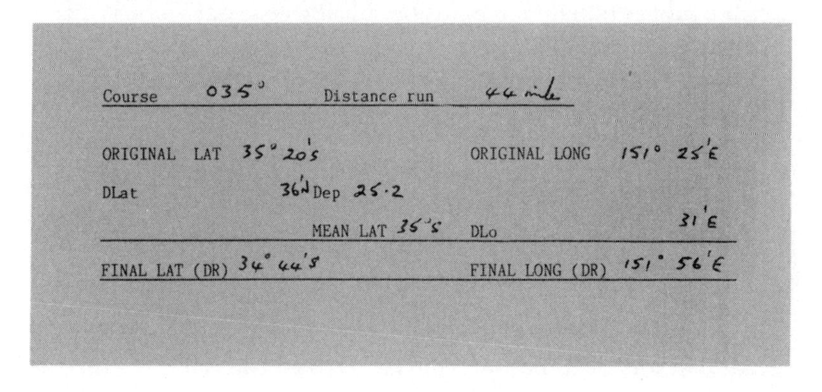

Course 035° Distance run 44 mile

ORIGINAL LAT 35° 20′S ORIGINAL LONG 151° 25′E

DLat 36′N Dep 25.2

MEAN LAT 35°S DLo 31′E

FINAL LAT (DR) 34° 44′S FINAL LONG (DR) 151° 56′E

Chapter 5 **The Celestial Globe**

In celestial navigation, the earth is assumed to be a relatively small globe completely surrounded by a huge 'plastic' globe on which all the heavenly bodies are located. This is not true of course; the sun, moon and planets are relatively close to earth compared to stars which may be millions of light years away. But from a navigational viewpoint the plastic celestial globe is a convenient arrangement upon which to base computations.

In order to locate the position of objects on earth a grid of latitude and longitude is used. An identical grid is used to locate heavenly bodies on the celestial globe, matching exactly the latitude and longitude on earth. The only difference between them is in name, since latitude on the celestial globe is termed *declination* while longitude is termed *Greenwich hour angle*, or more commonly just GHA.

Declination (Dec)

Like latitude, declination begins at zero on the celestial equator and runs north and south to the poles. It is measured in degrees and minutes and named N or S to indicate which hemisphere it is in.

Greenwich hour angle (GHA)

Longitude on earth begins at the Greenwich or *prime meridian* (0°). GHA commences on an identical meridian—the *celestial meridian* (0°). However, where longitude is measured 180° East and West of Greenwich, GHA is measured 360° right round the globe in a westwards direction. Thus Long 90°W would have its equivalent as GHA 90° on the celestial sphere. Long 180°E or W would be GHA 180°. But Long 90°E would have as its equivalent on the celesial sphere a GHA of 270°.

To illuminate the relationship between latitude and longitude on earth and declination and GHA on the celestial globe we can

The celestial globe.

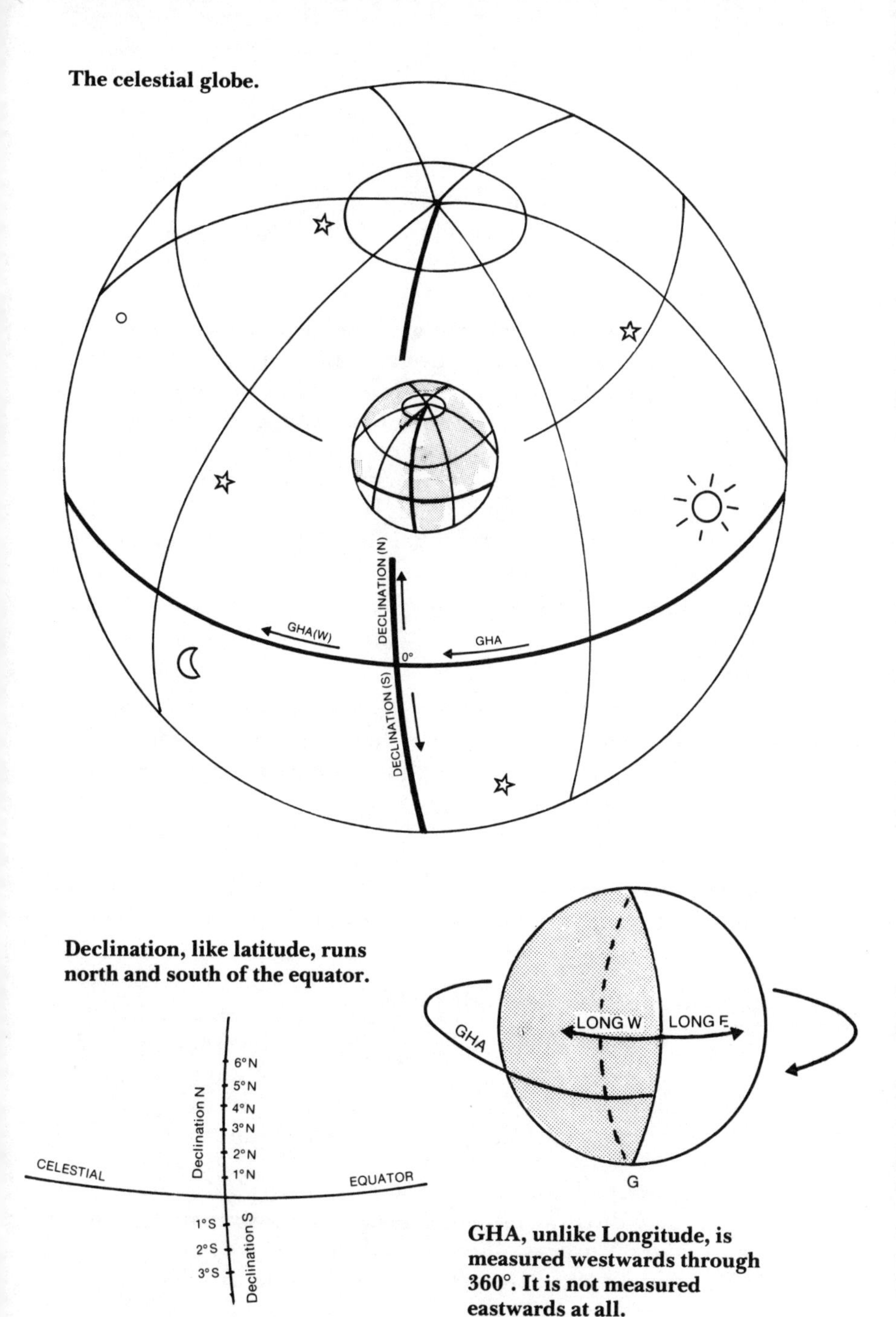

Declination, like latitude, runs north and south of the equator.

GHA, unlike Longitude, is measured westwards through 360°. It is not measured eastwards at all.

use the sun as an example. If you were standing on earth in Lat 20°N, Long 100°W and the sun's position were Dec 20°N, GHA 100° the sun would be immediately overhead and your body would throw no shadow. A similar example would be as follows:

Boat's position: Lat 16°50′S Long 145°30′E
Sun overhead (no shadows) Dec 16°50′S GHA 214°30′

Local hour angle (LHA)

The measurement of an hour angle is defined as the number of degrees and minutes between two meridians. Thus GHA is the angle between the Greenwich meridian and the meridian on which the heavenly body is located. LHA is the angle between the boat's meridian and the meridian on which the heavenly body is located. Thus if the boat is situated on the longitude meridian 90°W and the sun is rising with a GHA of 60°, then the LHA is 30°E. From this results a formula which is used frequently in celestial navigation:

Longitude GHA±LHA.

'Aries' (♈)

The stars are outside our solar system and therefore must be treated somewhat differently from the sun, moon and planets. For this reason a 'jumping off' meridian is selected on the celestial globe from which the hour angle of all stars is measured. This is termed the *first point of Aries,* more commonly called just 'Aries'. It is the reference point on the celestial globe for all stars in the outer universe and has a GHA which is the angle from the Greenwich meridian to the meridian on which Aries is located.

Sidereal (star's) hour angle (SHA)

This is the measurement in degrees and minutes from the meridian of Aries to the meridian on which the star is located. It is fairly constant, changing little in a year. It is added to the GHA of Aries to obtain the GHA of the star. The formula then reads:

GHA Star = GHA Aries + SHA Star.

Lets liken it to measuring the distance between, say, London, England and Newport, Rhode Is. It is unlikely any table will list the distance direct, so the best way would be to first find the distance between London and New York, and add it to the dis-

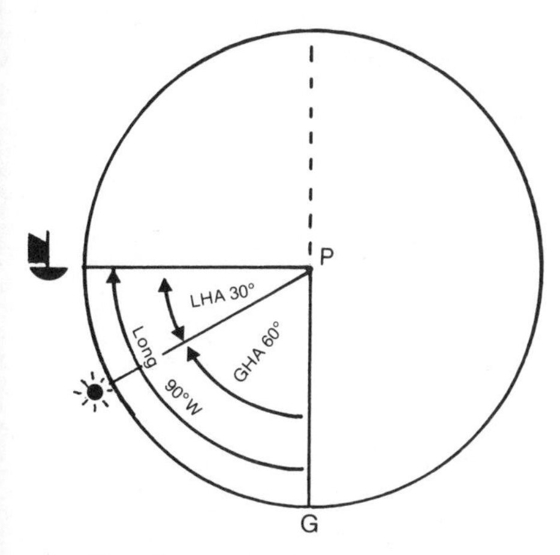

The diagram assumes we are looking down on the world from a point over the North Pole (P) with the equator as the circumference. The Greenwich meridian is represented by PG and the respective meridians of sun and boat as indicated.

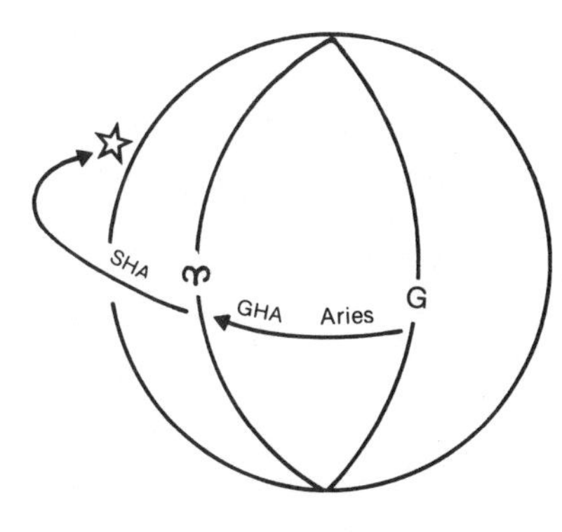

Aries, and its relationship to a star.

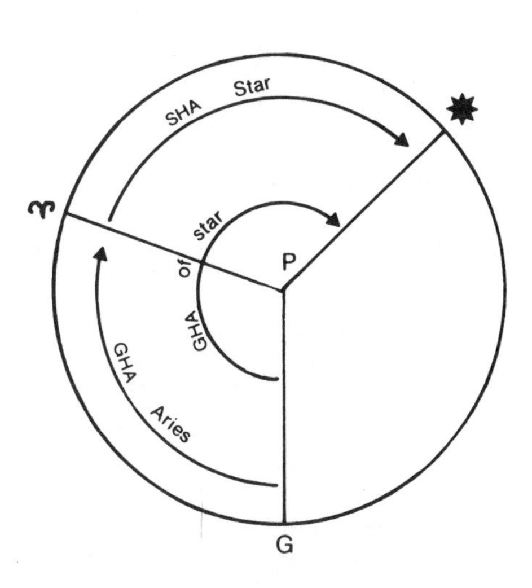

The relationship between Aries and GHA and SHA.

tance from New York to Newport, Rhode Is. Similarly, the easiest way to find the GHA of a star is to first find the GHA of Aries and add it to the HA from Aries to the star (SHA).

Hour angle diagram

Although you can use formulae for hour angle computation, a much simpler method is to draw what is known as the Hour Angle Diagram. Providing this diagram is correctly drawn and the meridians correctly placed, the solution to each computation becomes self-evident. The diagram is drawn in the plane of the

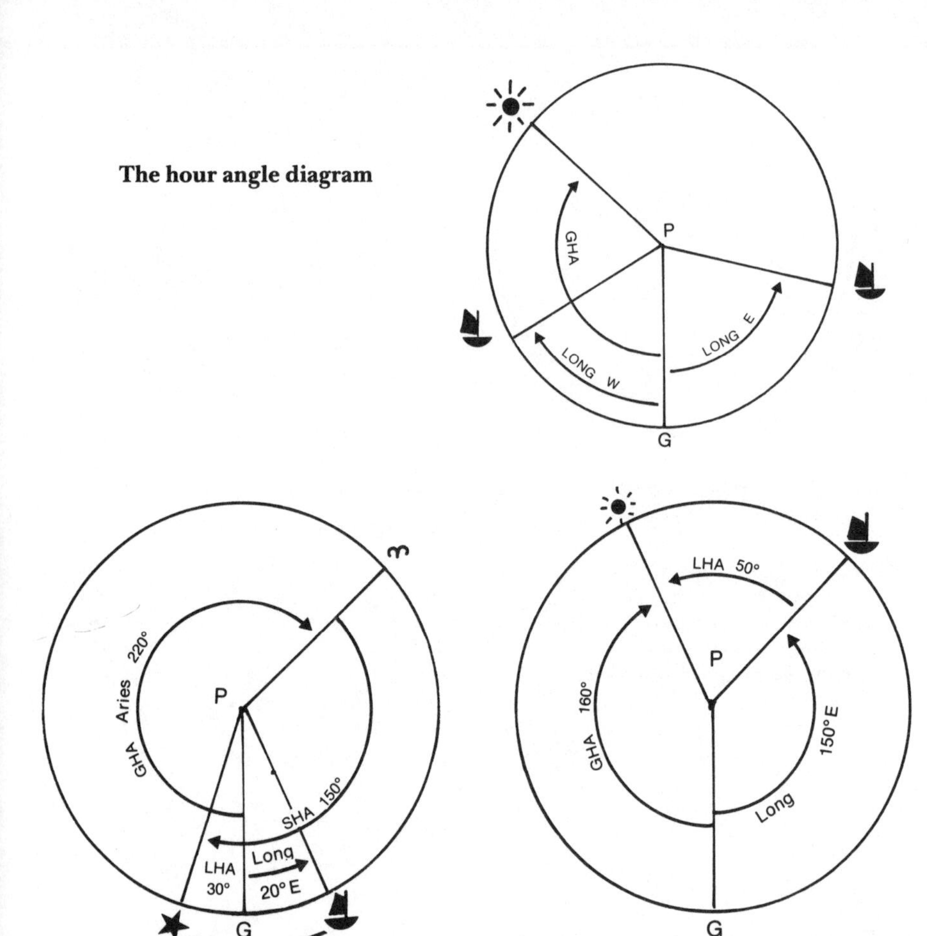

The hour angle diagram

A more complicated example of the use of the hour angle diagram with a star.

The hour angle diagram using the sun.

equator looking down on the North Pole and with the Greenwich meridian at the '6 o'clock' position. Since all hour angles are measured between meridians, the meridians of Greenwich, the boat and the heavenly body being used can each be represented by a radius of the circle, the center of which is the North Pole.

Time

As already explained in the section on the taking of a sight, time is of the utmost importance in celestial navigation. There are two forms of time on board a boat, GMT (Greenwich Mean Time) which is used for all navigational problems, and LMT (Local Mean Time) which is the normal day-to-day boat's time. LMT is used only for getting up, going to bed, meals and watch-keeping! It has no relationship to navigation. So it is common practice on board a boat to keep two clocks going, one on GMT for navigational purposes and one on LMT for day-to-day activities.

There are other forms of time but these are usually local. Zone Time, for example, is purely a commercial convenience in which developed areas of the world are grouped together. If this were not so, and time were taken by the sun as it is at sea, then every village and town across a continent would have different times, varying perhaps by minutes and seconds. To prevent this, one-hour zones, sometimes even half-hour zones, are created in order to group together a number of towns or areas.

The 24-hour clock

Time, in navigation, is expressed in four figures or, as it is often known, the 24-hour clock. Time signals and other communications at sea are listed in this way and therefore it is important if you plan to navigate internationally to use the 24-hour clock. It is a sensible form of time and has much to recommend it. The following indicates the 24-hour clock arrangement:

HOUR	AM	HOUR	PM
0100	1	1300	1
0200	2	1400	2
0300	3	1500	3
0400	4	1600	4
0500	5	1700	5
0600	6	1800	6
0700	7	1900	7
0800	8	2000	8
0900	9	2100	9
1000	10	2200	10
1100	11	2300	11
1200	noon	2400	midnight

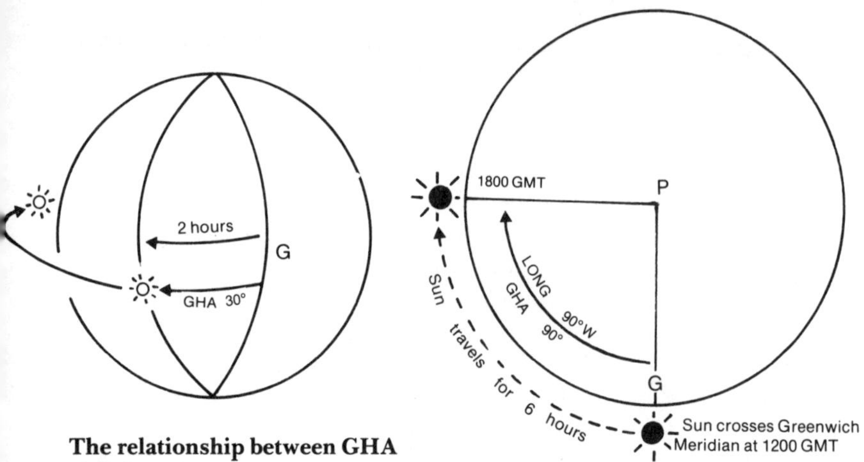

The relationship between GHA and time.

Illustrating the relationship between GHA and GMT.

Time and longitude

The sun travels around the world from east to west and covers 360° of longitude each 24 hours. Thus, at any given time throughout the day, it is possible to find the sun's longitude (or GHA) simply by converting the number of hours it has travelled from Greenwich into degrees and minutes of arc. The sun crosses the Greenwich meridian (0° longitude and GHA) at noon (1200 hours GMT) and thus all measurements are made from this point. For example, at 1800 hours GMT the sun will be in Long 90°W and the GHA will also be 90°.

Meridian passage of the sun (noon)

Noon on board a ship (1200 Hours LMT) is determined by the sun crossing the boat's meridian. In terms of the sextant sight, the sun reaches its *zenith*; that is, it stops rising from the east as it has done all morning and begins to fall towards setting in the west. This can be seen quite distinctly in the sextant, and indeed the noon sight is taken visually as the sun 'peaks'.

A practical application of the time/longitude relationship is to find the GMT of the noon 'meridian passage' as it is called. Since the boat's clocks were set the previous noon the boat has travelled over a certain amount of longitude, so there will be a discrepancy in her time. When travelling by ship or aircraft this time discrepancy is adjusted periodically by advancing or retarding watches to the next LMT. On board a boat at sea the same procedure must be followed, except that clocks are usually adjusted at noon.

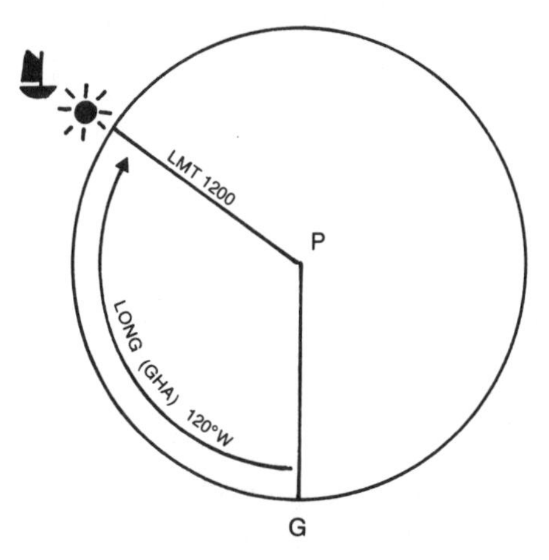

In order to find the time of the sun's meridian passage, the noon DR longitude is converted into time and applied to noon GMT. Thus, if the boat is in Long 120°W the sun will take eight hours to travel from the Greenwich meridian to the boat's meridian. It crosses the Greenwich meridian at 1200 GMT, therefore the time it crosses the boat's meridian (1200 LMT) will be eight hours later, i.e. 2000 hours GMT.

The time diagram

This is identical to the hour angle diagram and works on the same principle. Indeed, the two can be combined, since, as described, there is a close relationship between longitude, GHA and time. The example given in the previous paragraph for finding time of noon in Long 120°W is typical of problems which can be easily resolved by the combined time/hour angle diagram.

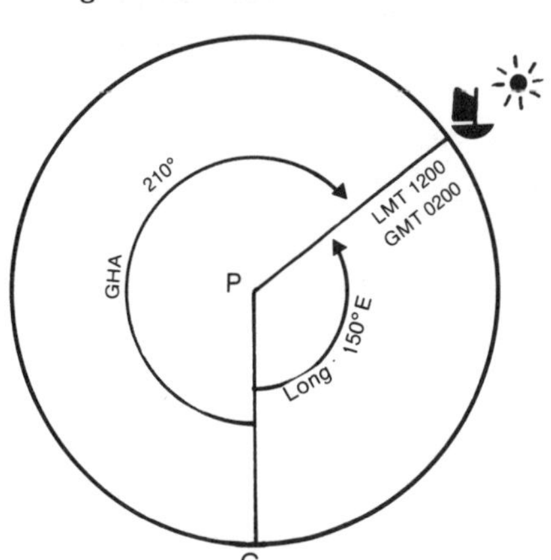

The time diagram, illustrating the relationship between hour angle, longitude and time.

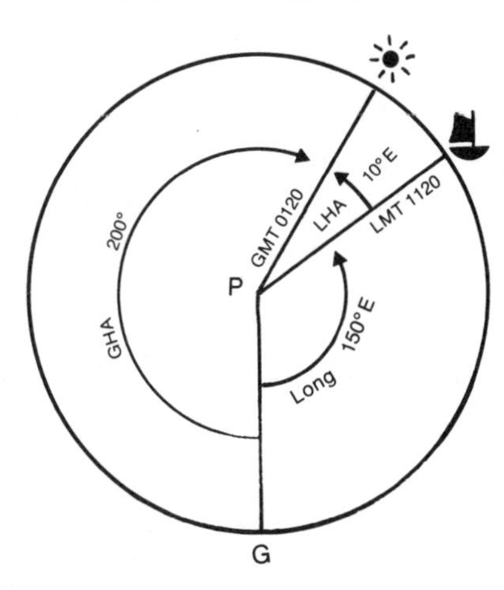

A more complicated example of the use of the time diagram.

Chapter 6 **The tables**

When a sight is taken, two sets of figures are obtained—the sextant altitude read directly from the instrument, and the GMT of the sight read from the chronometer. Before these can be used in any sight computation, both must be corrected—the GMT for any errors in the chronometer, and the altitude for the various sight corrections as described in Chapter 3. With this done the results will be processed through one of a number of methods of sight computation which will ultimately result in a plot of the boat's position.

Since this is a book for beginners, I have adopted the two simplest forms of position fixing which, despite their simplicity, are nonetheless very accurate. Indeed, navigators on naval and commercial ships use these methods as everyday navigational routine. While there are other methods of resolving celestial sights, they are generally more involved mathematically and are not widely used these days.

In addition, the use of these two methods of sight reductions require only two volumes of tables—*The Nautical Almanac* and the *Sight Reduction Tables for Marine Navigation* (DMAHTC Publication 229). Space, weight and cost are important factors to be considered when setting off on an ocean passage, so keeping navigation volumes to only two is a good practice.

The methods adopted here for the navigator's day's work are:

Dawn star sights Full fix position by sight reduction method and plotting.

Morning sun sight Position line by sight reduction method.

Noon sight Latitude by meridian altitude method crossed with morning position line.

Afternoon sun sight Position line by sight reduction method crossed with noon latitude.

Dusk star sights Full fix position by sight reduction method and plotting.

SIGHT REDUCTION FORM

SUN/STAR/PLANET DATE OF SIGHT LMT

GMT d h m s DR LAT DR LONG

Sextant

Sext alt
I.E. + –
Obs alt
Dip –
App Alt
Alt Corr – +
TRUE ALT (Ho)

Almanac

GHA (Aries for stars)
Inc +
SHA + (Stars only)
Total GHA
a LON – +
Tab LHA

TAB DEC DEC INC a LAT

Tables (HO 229)

Hc d Z

corr (1).......
corr (2).......

Tot corr >>>>>> (+ – according to d)

Hc

Ho

INTERCEPT (T/A) AZIMUTH (Zn)

NOON LATITUDE

Sext alt
I.E. – +
Obs alt
Dip –
App alt
Alt Corr +
Ho

90° 00'
Ho –
ZX
Dec + –

NOON LAT

The sight reduction and noon latitude form.

242 1984 DECEMBER 14, 15, 16 (FRI., SAT., SUN.)

G.M.T.	ARIES G.H.A.	VENUS −3.7 G.H.A.	VENUS Dec.	MARS +1.0 G.H.A.	MARS Dec.	JUPITER −1.5 G.H.A.	JUPITER Dec.	SATURN +0.8 G.H.A.	SATURN Dec.	STARS Name	S.H.A.	Dec.
d h	° ′	° ′	° ′	° ′	° ′	° ′	° ′	° ′	° ′		° ′	° ′
14 00	82 50.8	134 11.8	S21 00.7	118 40.0	S15 32.8	154 00.0	S22 32.4	211 49.7	S16 30.7	Acamar	315 34.4	S40 22.0
01	97 53.2	149 11.2	20 59.9	133 40.6	32.1	169 01.9	32.3	226 51.9	30.8	Achernar	335 42.4	S57 19.1
02	112 55.7	164 10.7	59.2	148 41.2	31.5	184 03.7	32.2	241 54.0	30.8	Acrux	173 34.2	S63 00.6
03	127 58.2	179 10.1 ··	58.4	163 41.8 ··	30.8	199 05.6 ··	32.2	256 56.2 ··	30.9	Adhara	255 29.3	S28 56.9
04	143 00.6	194 09.6	57.6	178 42.4	30.2	214 07.5	32.1	271 58.4	31.0	Aldebaran	291 14.1	N16 28.9
05	158 03.1	209 09.0	56.8	193 43.0	29.5	229 09.3	32.0	287 00.6	31.0			
06	173 05.5	224 08.5	S20 56.1	208 43.6	S15 28.8	244 11.2	S22 32.0	302 02.8	S16 31.1	Alioth	166 39.7	N56 02.3
07	188 08.0	239 07.9	55.3	223 44.2	28.2	259 13.1	31.9	317 05.0	31.2	Alkaid	153 16.2	N49 23.1
08	203 10.5	254 07.4	54.5	238 44.8	27.5	274 14.9	31.8	332 07.2	31.2	Al Na'ir	28 11.0	S47 02.4
F 09	218 12.9	269 06.8 ··	53.7	253 45.4 ··	26.9	289 16.8 ··	31.8	347 09.4 ··	31.3	Alnilam	276 08.1	S 1 12.6
R 10	233 15.4	284 06.3	53.0	268 46.0	26.2	304 18.7	31.7	2 11.5	31.4	Alphard	218 17.3	S 8 35.4
I 11	248 17.9	299 05.7	52.2	283 46.6	25.6	319 20.5	31.6	17 13.7	31.4			
D 12	263 20.3	314 05.2	S20 51.4	298 47.2	S15 24.9	334 22.4	S22 31.6	32 15.9	S16 31.5	Alphecca	126 29.7	N26 45.8
A 13	278 22.8	329 04.6	50.6	313 47.8	24.3	349 24.3	31.5	47 18.1	31.6	Alpheratz	358 06.1	N29 00.5
Y 14	293 25.3	344 04.1	49.8	328 48.4	23.6	4 26.1	31.4	62 20.3	31.6	Altair	62 29.8	N 8 49.6
15	308 27.7	359 03.6 ··	49.1	343 49.1 ··	23.0	19 28.0 ··	31.4	77 22.5 ··	31.7	Ankaa	353 36.9	S42 23.6
16	323 30.2	14 03.0	48.3	358 49.7	22.3	34 29.9	31.3	92 24.7	31.8	Antares	112 53.4	S26 23.9
17	338 32.6	29 02.5	47.5	13 50.3	21.6	49 31.7	31.2	107 26.9	31.8			
18	353 35.1	44 01.9	S20 46.7	28 50.9	S15 21.0	64 33.6	S22 31.2	122 29.0	S16 31.9	Arcturus	146 15.8	N19 15.6
19	8 37.6	59 01.4	45.9	43 51.5	20.3	79 35.5	31.1	137 31.2	31.9	Atria	108 15.5	S69 00.1
20	23 40.0	74 00.9	45.1	58 52.1	19.7	94 37.3	31.0	152 33.4	32.0	Avior	234 26.6	S59 27.3
21	38 42.5	89 00.3 ··	44.3	73 52.7 ··	19.0	109 39.2 ··	30.9	167 35.6 ··	32.1	Bellatrix	278 55.1	N 6 20.3
22	53 45.0	103 59.8	43.6	88 53.3	18.4	124 41.1	30.9	182 37.8	32.1	Betelgeuse	271 24.5	N 7 24.4
23	68 47.4	118 59.3	42.8	103 53.9	17.7	139 42.9	30.8	197 40.0	32.2			
15 00	83 49.9	133 58.7	S20 42.0	118 54.5	S15 17.1	154 44.8	S22 30.7	212 42.2	S16 32.3	Canopus	264 05.3	S52 41.1
01	98 52.4	148 58.2	41.2	133 55.1	16.4	169 46.6	30.7	227 44.4	32.3	Capella	281 06.2	N45 59.1
02	113 54.8	163 57.7	40.4	148 55.7	15.7	184 48.5	30.6	242 46.6	32.4	Deneb	49 46.8	N45 13.6
03	128 57.3	178 57.1 ··	39.6	163 56.3 ··	15.1	199 50.4 ··	30.5	257 48.7 ··	32.5	Denebola	182 55.8	N14 39.4
04	143 59.8	193 56.6	38.8	178 57.0	14.4	214 52.2	30.5	272 50.9	32.5	Diphda	349 17.6	S18 04.3
05	159 02.2	208 56.1	38.0	193 57.6	13.8	229 54.1	30.4	287 53.1	32.6			
06	174 04.7	223 55.5	S20 37.2	208 58.2	S15 13.1	244 56.0	S22 30.3	302 55.3	S16 32.7	Dubhe	194 17.9	N61 49.8
07	189 07.1	238 55.0	36.4	223 58.8	12.4	259 57.8	30.3	317 57.5	32.7	Elnath	278 39.8	N28 35.8
S 08	204 09.6	253 54.5	35.6	238 59.4	11.8	274 59.7	30.2	332 59.7	32.8	Eltanin	90 56.8	N51 29.4
A 09	219 12.1	268 53.9 ··	34.8	254 00.0 ··	11.1	290 01.6 ··	30.1	348 01.9 ··	32.9	Enif	34 08.7	N 9 48.3
T 10	234 14.5	283 53.4	34.0	269 00.6	10.5	305 03.4	30.0	3 04.1	32.9	Fomalhaut	15 47.9	S29 42.4
U 11	249 17.0	298 52.9	33.2	284 01.2	09.8	320 05.3	30.0	18 06.3	33.0			
R 12	264 19.5	313 52.4	S20 32.4	299 01.8	S15 09.1	335 07.2	S22 29.9	33 08.4	S16 33.1	Gacrux	172 25.6	S57 01.4
D 13	279 21.9	328 51.8	31.6	314 02.4	08.5	350 09.0	29.8	48 10.6	33.1	Gienah	176 14.9	S17 27.3
A 14	294 24.4	343 51.3	30.8	329 03.0	07.8	5 10.9	29.8	63 12.8	33.2	Hadar	149 19.6	S60 17.8
Y 15	309 26.9	358 50.8 ··	30.0	344 03.7 ··	07.2	20 12.8 ··	29.7	78 15.0 ··	33.3	Hamal	328 25.2	N23 23.6
16	324 29.3	13 50.3	29.2	359 04.3	06.5	35 14.6	29.6	93 17.2	33.3	Kaus Aust.	84 13.1	S34 23.7
17	339 31.8	28 49.8	28.4	14 04.9	05.8	50 16.5	29.6	108 19.4	33.4			
18	354 34.3	43 49.2	S20 27.6	29 05.5	S15 05.2	65 18.4	S22 29.5	123 21.6	S16 33.4	Kochab	137 19.7	N74 12.8
19	9 36.7	58 48.7	26.8	44 06.1	04.5	80 20.2	29.4	138 23.8	33.5	Markab	14 00.1	N15 07.4
20	24 39.2	73 48.2	26.0	59 06.7	03.9	95 22.1	29.3	153 26.0	33.6	Menkar	314 37.6	N 4 01.9
21	39 41.6	88 47.7 ··	25.1	74 07.3 ··	03.2	110 23.9 ··	29.3	168 28.1 ··	33.6	Menkent	148 33.7	S36 17.6
22	54 44.1	103 47.2	24.3	89 07.9	02.5	125 25.8	29.2	183 30.3	33.7	Miaplacidus	221 44.0	S69 38.9
23	69 46.6	118 46.6	23.5	104 08.5	01.9	140 27.7	29.1	198 32.5	33.8			
16 00	84 49.0	133 46.1	S20 22.7	119 09.2	S15 01.2	155 29.5	S22 29.1	213 34.7	S16 33.8	Mirfak	309 11.3	N49 48.7
01	99 51.5	148 45.6	21.9	134 09.8	15 00.5	170 31.4	29.0	228 36.9	33.9	Nunki	76 25.6	S26 19.1
02	114 54.0	163 45.1	21.1	149 10.4	14 59.9	185 33.3	28.9	243 39.1	34.0	Peacock	53 53.8	S56 47.4
03	129 56.4	178 44.6 ··	20.3	164 11.0 ··	59.2	200 35.1 ··	28.9	258 41.3 ··	34.0	Pollux	243 54.0	N28 03.8
04	144 58.9	193 44.1	19.4	179 11.6	58.5	215 37.0	28.8	273 43.5	34.1	Procyon	245 22.2	N 5 16.0
05	160 01.4	208 43.6	18.6	194 12.2	57.9	230 38.9	28.7	288 45.7	34.2			
06	175 03.8	223 43.0	S20 17.8	209 12.8	S14 57.2	245 40.7	S22 28.6	303 47.9	S16 34.2	Rasalhague	96 27.0	N12 34.2
07	190 06.3	238 42.5	17.0	224 13.4	56.6	260 42.6	28.6	318 50.0	34.3	Regulus	208 06.5	N12 02.5
08	205 08.8	253 42.0	16.2	239 14.0	55.9	275 44.4	28.5	333 52.2	34.4	Rigel	281 32.7	S 8 13.0
S 09	220 11.2	268 41.5 ··	15.3	254 14.7 ··	55.2	290 46.3 ··	28.4	348 54.4 ··	34.4	Rigil Kent.	140 22.3	S60 46.1
U 10	235 13.7	283 41.0	14.5	269 15.3	54.6	305 48.2	28.4	3 56.6	34.5	Sabik	102 37.9	S15 42.5
N 11	250 16.1	298 40.5	13.7	284 15.9	53.9	320 50.0	28.3	18 58.8	34.5			
D 12	265 18.6	313 40.0	S20 12.9	299 16.5	S14 53.2	335 51.9	S22 28.2	34 01.0	S16 34.6	Schedar	350 05.5	N56 27.5
A 13	280 21.1	328 39.5	12.1	314 17.1	52.6	350 53.8	28.2	49 03.2	34.7	Shaula	96 52.0	S37 05.7
Y 14	295 23.5	343 39.0	11.2	329 17.7	51.9	5 55.6	28.1	64 05.4	34.7	Sirius	258 52.6	S16 41.6
15	310 26.0	358 38.5 ··	10.4	344 18.3 ··	51.2	20 57.5 ··	28.0	79 07.6 ··	34.8	Spica	158 54.4	S11 04.9
16	325 28.5	13 38.0	09.6	359 19.0	50.6	35 59.4	27.9	94 09.8	34.9	Suhail	223 08.3	S43 22.0
17	340 30.9	28 37.5	08.7	14 19.6	49.9	51 01.2	27.9	109 12.0	34.9			
18	355 33.4	43 37.0	S20 07.9	29 20.2	S14 49.2	66 03.1	S22 27.8	124 14.1	S16 35.0	Vega	80 54.2	N38 46.1
19	10 35.9	58 36.5	07.1	44 20.8	48.6	81 04.9	27.7	139 16.3	35.1	Zuben'ubi	137 29.9	S15 58.7
20	25 38.3	73 36.0	06.3	59 21.4	47.9	96 06.8	27.7	154 18.5	35.1		S.H.A.	Mer. Pass.
21	40 40.8	88 35.5 ··	05.4	74 22.0 ··	47.2	111 08.7 ··	27.6	169 20.7 ··	35.2		° ′	h m
22	55 43.2	103 35.0	04.6	89 22.6	46.5	126 10.5	27.5	184 22.9	35.3	Venus	50 08.8	15 05
23	70 45.7	118 34.5	03.8	104 23.3	45.9	141 12.4	27.4	199 25.1	35.3	Mars	35 04.6	16 04
										Jupiter	70 54.9	13 39
Mer. Pass.	18 21.7	v −0.5	d 0.8	v 0.6	d 0.7	v 1.9	d 0.1	v 2.2	d 0.1	Saturn	128 52.3	9 48

An excerpt from the daily pages of *The Nautical Almanac*.

The sight form

I have devised one sight form to cover both sight reduction and meridian altitude method of sights. This is the only form required for a complete day's navigational calculations using these two methods.

The Nautical Almanac

The Nautical Almanac jointly published by the U.S. Naval Observatory and HM Nautical Almanac Office at the Royal Greenwich Observatory, lists the positions of all heavenly bodies for every second of every day of the year. These positions are listed

G.M.T.	SUN G.H.A.	SUN Dec.	MOON G.H.A.	v	MOON Dec.	d	H.P.
d h	° ′	° ′	° ′	′	° ′	′	′
14 00	181 21.6	S23 12.7	288 03.1	9.9	N16 03.8	12.8	58.6
01	196 21.3	12.9	302 32.0	10.0	15 51.0	12.8	58.6
02	211 21.0	13.0	317 01.0	10.1	15 38.2	13.0	58.6
03	226 20.7	·· 13.2	331 30.1	10.1	15 25.2	13.0	58.6
04	241 20.4	13.3	345 59.2	10.2	15 12.2	13.1	58.7
05	256 20.1	13.4	0 28.4	10.2	14 59.1	13.2	58.7
06	271 19.8	S23 13.6	14 57.6	10.3	N14 45.9	13.3	58.7
07	286 19.5	13.7	29 26.9	10.4	14 32.6	13.4	58.7
08	301 19.2	13.9	43 56.3	10.4	14 19.2	13.4	58.7
F 09	316 18.9	·· 14.0	58 25.7	10.4	14 05.8	13.5	58.7
R 10	331 18.6	14.2	72 55.1	10.5	13 52.3	13.6	58.7
I 11	346 18.3	14.3	87 24.6	10.5	13 38.7	13.7	58.8
D 12	1 18.0	S23 14.4	101 54.1	10.6	N13 25.0	13.8	58.8
A 13	16 17.7	14.6	116 23.7	10.7	13 11.2	13.8	58.8
Y 14	31 17.4	14.7	130 53.4	10.6	12 57.4	13.9	58.8
15	46 17.1	·· 14.9	145 23.0	10.8	12 43.5	14.0	58.8
16	61 16.8	15.0	159 52.8	10.7	12 29.5	14.1	58.8
17	76 16.5	15.1	174 22.5	10.9	12 15.4	14.1	58.8
18	91 16.2	S23 15.3	188 52.4	10.8	N12 01.3	14.2	58.9
19	106 15.9	15.4	203 22.2	10.9	11 47.1	14.3	58.9
20	121 15.6	15.5	217 52.1	11.0	11 32.8	14.3	58.9
21	136 15.3	·· 15.7	232 22.1	11.0	11 18.5	14.4	58.9
22	151 15.0	15.8	246 52.1	11.0	11 04.1	14.5	58.9
23	166 14.7	15.9	261 22.1	11.1	10 49.6	14.5	58.9
15 00	181 14.4	S23 16.1	275 52.2	11.1	N10 35.1	14.6	58.9
01	196 14.1	16.2	290 22.3	11.1	10 20.5	14.6	58.9
02	211 13.8	16.3	304 52.4	11.2	10 05.9	14.7	59.0
03	226 13.5	·· 16.5	319 22.6	11.2	9 51.2	14.8	59.0
04	241 13.2	16.6	333 52.8	11.3	9 36.4	14.8	59.0
05	256 12.9	16.7	348 23.1	11.3	9 21.6	14.9	59.0
06	271 12.6	S23 16.8	2 53.4	11.3	N 9 06.7	14.9	59.0
07	286 12.3	17.0	17 23.7	11.3	8 51.8	14.9	59.0
S 08	301 12.0	17.1	31 54.0	11.4	8 36.9	15.1	59.0
A 09	316 11.7	·· 17.2	46 24.4	11.4	8 21.8	15.0	59.0
T 10	331 11.4	17.3	60 54.8	11.5	8 06.8	15.2	59.0
U 11	346 11.1	17.5	75 25.3	11.4	7 51.6	15.1	59.1
R 12	1 10.8	S23 17.6	89 55.7	11.5	N 7 36.5	15.2	59.1
D 13	16 10.5	17.7	104 26.2	11.5	7 21.3	15.3	59.1
A 14	31 10.2	17.8	118 56.7	11.6	7 06.0	15.3	59.1
Y 15	46 09.9	·· 17.9	133 27.3	11.5	6 50.7	15.3	59.1
16	61 09.5	18.1	147 57.8	11.6	6 35.4	15.4	59.1
17	76 09.2	18.2	162 28.4	11.6	6 20.0	15.4	59.1
18	91 08.9	S23 18.3	176 59.0	11.6	N 6 04.6	15.5	59.1
19	106 08.6	18.4	191 29.6	11.6	5 49.1	15.4	59.1
20	121 08.3	18.5	206 00.2	11.7	5 33.7	15.6	59.2
21	136 08.0	·· 18.6	220 30.9	11.7	5 18.1	15.5	59.2
22	151 07.7	18.7	235 01.6	11.6	5 02.6	15.6	59.2
23	166 07.4	18.9	249 32.2	11.7	4 47.0	15.6	59.2
16 00	181 07.1	S23 19.0	264 02.9	11.7	N 4 31.4	15.6	59.2
01	196 06.8	19.1	278 33.6	11.7	4 15.8	15.7	59.2
02	211 06.5	19.2	293 04.3	11.8	4 00.1	15.7	59.2
03	226 06.2	·· 19.3	307 35.1	11.7	3 44.4	15.7	59.2
04	241 05.9	19.4	322 05.8	11.7	3 28.7	15.8	59.2
05	256 05.6	19.5	336 36.5	11.8	3 12.9	15.7	59.2
06	271 05.3	S23 19.6	351 07.3	11.7	N 2 57.2	15.8	59.3
07	286 05.0	19.7	5 38.0	11.8	2 41.4	15.8	59.3
08	301 04.7	19.8	20 08.8	11.7	2 25.6	15.8	59.3
S 09	316 04.4	·· 19.9	34 39.5	11.8	2 09.8	15.9	59.3
U 10	331 04.1	20.0	49 10.3	11.8	1 53.9	15.8	59.3
N 11	346 03.8	20.1	63 41.1	11.7	1 38.1	15.9	59.3
D 12	1 03.5	S23 20.2	78 11.8	11.8	N 1 22.2	15.9	59.3
A 13	16 03.2	20.3	92 42.6	11.7	1 06.3	15.9	59.3
Y 14	31 02.9	20.4	107 13.3	11.7	0 50.4	15.9	59.3
15	46 02.6	·· 20.5	121 44.0	11.8	0 34.5	15.9	59.3
16	61 02.3	20.6	136 14.8	11.7	0 18.6	15.9	59.3
17	76 01.9	20.7	150 45.5	11.7	N 0 02.7	15.9	59.4
18	91 01.6	S23 20.8	165 16.2	11.7	S 0 13.2	15.9	59.4
19	106 01.3	20.9	179 46.9	11.7	0 29.1	15.9	59.4
20	121 01.0	21.0	194 17.6	11.7	0 45.0	15.9	59.4
21	136 00.7	·· 21.1	208 48.3	11.7	1 00.9	16.0	59.4
22	151 00.4	21.2	223 19.0	11.6	1 16.9	15.9	59.4
23	166 00.1	21.3	237 49.6	11.6	1 32.8	15.9	59.4
	S.D. 16.3	d 0.1	S.D. 16.0		16.1		16.2

Lat.	Twilight Naut.	Twilight Civil	Sunrise	Moonrise 14	Moonrise 15	Moonrise 16	Moonrise 17
°	h m	h m	h m	h m	h m	h m	h m
N 72	08 20	10 47	▬	21 01	23 27	25 44	01 44
N 70	08 01	09 48	▬	21 21	23 33	25 41	01 41
68	07 45	09 14	▬	21 37	23 39	25 38	01 38
66	07 32	08 49	10 28	21 49	23 43	25 35	01 35
64	07 21	08 29	09 47	21 59	23 47	25 33	01 33
62	07 12	08 13	09 19	22 08	23 50	25 32	01 32
60	07 03	08 00	08 57	22 16	23 53	25 30	01 30
N 58	06 56	07 48	08 40	22 22	23 56	25 29	01 29
56	06 49	07 38	08 25	22 28	23 58	25 28	01 28
54	06 43	07 29	08 13	22 33	24 00	00 00	01 27
52	06 37	07 21	08 02	22 38	24 02	00 02	01 26
50	06 32	07 14	07 52	22 42	24 03	00 03	01 25
45	06 20	06 58	07 31	22 51	24 07	00 07	01 23
N 40	06 10	06 44	07 15	22 58	24 10	00 10	01 21
35	06 01	06 33	07 01	23 05	24 12	00 12	01 20
30	05 52	06 22	06 48	23 10	24 15	00 15	01 19
20	05 36	06 03	06 27	23 20	24 19	00 19	01 17
N 10	05 20	05 46	06 09	23 28	24 22	00 22	01 15
0	05 03	05 29	05 52	23 36	24 25	00 25	01 14
S 10	04 44	05 11	05 34	23 44	24 28	00 28	01 12
20	04 21	04 51	05 15	23 53	24 32	00 32	01 11
30	03 53	04 26	04 53	24 02	00 02	00 36	01 09
35	03 34	04 11	04 40	24 08	00 08	00 38	01 08
40	03 11	03 53	04 25	24 14	00 14	00 41	01 07
45	02 41	03 30	04 07	24 21	00 21	00 43	01 05
S 50	01 56	03 01	03 45	00 09	00 29	00 47	01 04
52	01 28	02 46	03 34	00 16	00 33	00 49	01 03
54	00 44	02 28	03 22	00 23	00 38	00 50	01 02
56	////	02 06	03 08	00 30	00 42	00 52	01 01
58	////	01 37	02 51	00 39	00 48	00 54	01 00
S 60	////	00 49	02 31	00 49	00 53	00 57	00 59

Lat.	Sunset	Twilight Civil	Twilight Naut.	Moonset 14	Moonset 15	Moonset 16	Moonset 17
°	h m	h m	h m	h m	h m	h m	h m
N 72	▬	13 03	15 30	14 14	13 33	13 01	12 29
N 70	▬	14 02	15 50	13 51	13 23	13 00	12 36
68	▬	14 37	16 05	13 34	13 15	12 58	12 42
66	13 23	15 02	16 18	13 20	13 08	12 58	12 47
64	14 04	15 21	16 29	13 08	13 02	12 57	12 51
62	14 32	15 37	16 39	12 58	12 57	12 56	12 55
60	14 53	15 50	16 47	12 49	12 53	12 56	12 58
N 58	15 10	16 02	16 55	12 41	12 49	12 55	13 01
56	15 25	16 12	17 01	12 34	12 45	12 55	13 04
54	15 38	16 21	17 07	12 28	12 42	12 54	13 06
52	15 49	16 29	17 13	12 23	12 39	12 54	13 08
50	15 58	16 37	17 18	12 18	12 36	12 53	13 10
45	16 19	16 53	17 30	12 07	12 31	12 53	13 15
N 40	16 36	17 06	17 40	11 58	12 26	12 52	13 18
35	16 50	17 18	17 50	11 50	12 21	12 51	13 21
30	17 02	17 28	17 58	11 43	12 18	12 51	13 24
20	17 23	17 47	18 15	11 31	12 11	12 50	13 29
N 10	17 42	18 04	18 31	11 20	12 05	12 49	13 33
0	17 59	18 22	18 48	11 10	12 00	12 48	13 37
S 10	18 17	18 40	19 07	11 00	11 54	12 47	13 41
20	18 35	19 00	19 29	10 49	11 48	12 46	13 45
30	18 57	19 25	19 58	10 36	11 41	12 45	13 50
35	19 10	19 40	20 17	10 29	11 37	12 45	13 53
40	19 25	19 58	20 40	10 21	11 32	12 44	13 56
45	19 43	20 21	21 10	10 11	11 27	12 43	13 59
S 50	20 06	20 50	21 55	09 59	11 21	12 42	14 04
52	20 16	21 05	22 23	09 53	11 18	12 41	14 06
54	20 29	21 23	23 08	09 47	11 14	12 41	14 08
56	20 43	21 45	////	09 40	11 11	12 40	14 11
58	20 59	22 14	////	09 32	11 06	12 40	14 13
S 60	21 20	23 03	////	09 23	11 02	12 39	14 16

Day	SUN Eqn. of Time 00h	SUN Eqn. of Time 12h	SUN Mer. Pass.	MOON Mer. Pass. Upper	MOON Mer. Pass. Lower	Age	Phase
	m s	m s	h m	h m	h m	d	
14	05 27	05 13	11 55	04 58	17 23	22	
15	04 58	04 44	11 55	05 48	18 12	23	◑
16	04 29	04 14	11 56	06 37	19 01	24	

An excerpt from the daily pages of *The Nautical Almanac.*

in GHA and Dec except in the case of stars where they are listed separately as Aries and SHA. In the main pages the positions are listed for every hour of GMT while a table of increments and corrections (tinted pages) at the back of the almanac allows interpolations for minutes and seconds. The main pages are in the form of double spreads, each of which contains three days data for sun, moon, stars and planets. In addition, tables relating to twilight, sunrise and sunset, moonrise and moonset are included.

I do not propose to deal with moon sights in this book, since of all celestial bodies, the moon is the most difficult. There is a close relationship between the sun and the planets, and stars are also

0m INCREMENTS AND CORRECTIONS

0m	SUN PLANETS	ARIES	MOON	v or d	Corrn	v or d	Corrn	v or d	Corrn
s	° ′	° ′	° ′	′	′	′	′	′	′
00	0 00·0	0 00·0	0 00·0	0·0	0·0	6·0	0·1	12·0	0·1
01	0 00·3	0 00·3	0 00·2	0·1	0·0	6·1	0·1	12·1	0·1
02	0 00·5	0 00·5	0 00·5	0·2	0·0	6·2	0·1	12·2	0·1
03	0 00·8	0 00·8	0 00·7	0·3	0·0	6·3	0·1	12·3	0·1
04	0 01·0	0 01·0	0 01·0	0·4	0·0	6·4	0·1	12·4	0·1
05	0 01·3	0 01·3	0 01·2	0·5	0·0	6·5	0·1	12·5	0·1
06	0 01·5	0 01·5	0 01·4	0·6	0·0	6·6	0·1	12·6	0·1
07	0 01·8	0 01·8	0 01·7	0·7	0·0	6·7	0·1	12·7	0·1
08	0 02·0	0 02·0	0 01·9	0·8	0·0	6·8	0·1	12·8	0·1
09	0 02·3	0 02·3	0 02·1	0·9	0·0	6·9	0·1	12·9	0·1
10	0 02·5	0 02·5	0 02·4	1·0	0·0	7·0	0·1	13·0	0·1
11	0 02·8	0 02·8	0 02·6	1·1	0·0	7·1	0·1	13·1	0·1
12	0 03·0	0 03·0	0 02·9	1·2	0·0	7·2	0·1	13·2	0·1
13	0 03·3	0 03·3	0 03·1	1·3	0·0	7·3	0·1	13·3	0·1
14	0 03·5	0 03·5	0 03·3	1·4	0·0	7·4	0·1	13·4	0·1
15	0 03·8	0 03·8	0 03·6	1·5	0·0	7·5	0·1	13·5	0·1
16	0 04·0	0 04·0	0 03·8	1·6	0·0	7·6	0·1	13·6	0·1
17	0 04·3	0 04·3	0 04·1	1·7	0·0	7·7	0·1	13·7	0·1
18	0 04·5	0 04·5	0 04·3	1·8	0·0	7·8	0·1	13·8	0·1
19	0 04·8	0 04·8	0 04·5	1·9	0·0	7·9	0·1	13·9	0·1
20	0 05·0	0 05·0	0 04·8	2·0	0·0	8·0	0·1	14·0	0·1
21	0 05·3	0 05·3	0 05·0	2·1	0·0	8·1	0·1	14·1	0·1
22	0 05·5	0 05·5	0 05·2	2·2	0·0	8·2	0·1	14·2	0·1
23	0 05·8	0 05·8	0 05·5	2·3	0·0	8·3	0·1	14·3	0·1
24	0 06·0	0 06·0	0 05·7	2·4	0·0	8·4	0·1	14·4	0·1
25	0 06·3	0 06·3	0 06·0	2·5	0·0	8·5	0·1	14·5	0·1
26	0 06·5	0 06·5	0 06·2	2·6	0·0	8·6	0·1	14·6	0·1
27	0 06·8	0 06·8	0 06·4	2·7	0·0	8·7	0·1	14·7	0·1
28	0 07·0	0 07·0	0 06·7	2·8	0·0	8·8	0·1	14·8	0·1
29	0 07·3	0 07·3	0 06·9	2·9	0·0	8·9	0·1	14·9	0·1
30	0 07·5	0 07·5	0 07·2	3·0	0·0	9·0	0·1	15·0	0·1
31	0 07·8	0 07·8	0 07·4	3·1	0·0	9·1	0·1	15·1	0·1
32	0 08·0	0 08·0	0 07·6	3·2	0·0	9·2	0·1	15·2	0·1
33	0 08·3	0 08·3	0 07·9	3·3	0·0	9·3	0·1	15·3	0·1
34	0 08·5	0 08·5	0 08·1	3·4	0·0	9·4	0·1	15·4	0·1
35	0 08·8	0 08·8	0 08·4	3·5	0·0	9·5	0·1	15·5	0·1
36	0 09·0	0 09·0	0 08·6	3·6	0·0	9·6	0·1	15·6	0·1
37	0 09·3	0 09·3	0 08·8	3·7	0·0	9·7	0·1	15·7	0·1
38	0 09·5	0 09·5	0 09·1	3·8	0·0	9·8	0·1	15·8	0·1
39	0 09·8	0 09·8	0 09·3	3·9	0·0	9·9	0·1	15·9	0·1
40	0 10·0	0 10·0	0 09·5	4·0	0·0	10·0	0·1	16·0	0·1
41	0 10·3	0 10·3	0 09·8	4·1	0·0	10·1	0·1	16·1	0·1
42	0 10·5	0 10·5	0 10·0	4·2	0·0	10·2	0·1	16·2	0·1
43	0 10·8	0 10·8	0 10·3	4·3	0·0	10·3	0·1	16·3	0·1
44	0 11·0	0 11·0	0 10·5	4·4	0·0	10·4	0·1	16·4	0·1
45	0 11·3	0 11·3	0 10·7	4·5	0·0	10·5	0·1	16·5	0·1
46	0 11·5	0 11·5	0 11·0	4·6	0·0	10·6	0·1	16·6	0·1
47	0 11·8	0 11·8	0 11·2	4·7	0·0	10·7	0·1	16·7	0·1
48	0 12·0	0 12·0	0 11·5	4·8	0·0	10·8	0·1	16·8	0·1
49	0 12·3	0 12·3	0 11·7	4·9	0·0	10·9	0·1	16·9	0·1
50	0 12·5	0 12·5	0 11·9	5·0	0·0	11·0	0·1	17·0	0·1
51	0 12·8	0 12·8	0 12·2	5·1	0·0	11·1	0·1	17·1	0·1
52	0 13·0	0 13·0	0 12·4	5·2	0·0	11·2	0·1	17·2	0·1
53	0 13·3	0 13·3	0 12·6	5·3	0·0	11·3	0·1	17·3	0·1
54	0 13·5	0 13·5	0 12·9	5·4	0·0	11·4	0·1	17·4	0·1
55	0 13·8	0 13·8	0 13·1	5·5	0·0	11·5	0·1	17·5	0·1
56	0 14·0	0 14·0	0 13·4	5·6	0·0	11·6	0·1	17·6	0·1
57	0 14·3	0 14·3	0 13·6	5·7	0·0	11·7	0·1	17·7	0·1
58	0 14·5	0 14·5	0 13·8	5·8	0·0	11·8	0·1	17·8	0·1
59	0 14·8	0 14·8	0 14·1	5·9	0·0	11·9	0·1	17·9	0·1
60	0 15·0	0 15·0	0 14·3	6·0	0·1	12·0	0·1	18·0	0·2

1m

1m	SUN PLANETS	ARIES	MOON	v or d	Corrn	v or d	Corrn	v or d	Corrn
s	° ′	° ′	° ′	′	′	′	′	′	′
00	0 15·0	0 15·0	0 14·3	0·0	0·0	6·0	0·2	12·0	0·3
01	0 15·3	0 15·3	0 14·6	0·1	0·0	6·1	0·2	12·1	0·3
02	0 15·5	0 15·5	0 14·8	0·2	0·0	6·2	0·2	12·2	0·3
03	0 15·8	0 15·8	0 15·0	0·3	0·0	6·3	0·2	12·3	0·3
04	0 16·0	0 16·0	0 15·3	0·4	0·0	6·4	0·2	12·4	0·3
05	0 16·3	0 16·3	0 15·5	0·5	0·0	6·5	0·2	12·5	0·3
06	0 16·5	0 16·5	0 15·7	0·6	0·0	6·6	0·2	12·6	0·3
07	0 16·8	0 16·8	0 16·0	0·7	0·0	6·7	0·2	12·7	0·3
08	0 17·0	0 17·0	0 16·2	0·8	0·0	6·8	0·2	12·8	0·3
09	0 17·3	0 17·3	0 16·5	0·9	0·0	6·9	0·2	12·9	0·3
10	0 17·5	0 17·5	0 16·7	1·0	0·0	7·0	0·2	13·0	0·3
11	0 17·8	0 17·8	0 16·9	1·1	0·0	7·1	0·2	13·1	0·3
12	0 18·0	0 18·0	0 17·2	1·2	0·0	7·2	0·2	13·2	0·3
13	0 18·3	0 18·3	0 17·4	1·3	0·0	7·3	0·2	13·3	0·3
14	0 18·5	0 18·6	0 17·7	1·4	0·0	7·4	0·2	13·4	0·3
15	0 18·8	0 18·8	0 17·9	1·5	0·0	7·5	0·2	13·5	0·3
16	0 19·0	0 19·1	0 18·1	1·6	0·0	7·6	0·2	13·6	0·3
17	0 19·3	0 19·3	0 18·4	1·7	0·0	7·7	0·2	13·7	0·3
18	0 19·5	0 19·6	0 18·6	1·8	0·0	7·8	0·2	13·8	0·3
19	0 19·8	0 19·8	0 18·9	1·9	0·0	7·9	0·2	13·9	0·3
20	0 20·0	0 20·1	0 19·1	2·0	0·1	8·0	0·2	14·0	0·4
21	0 20·3	0 20·3	0 19·3	2·1	0·1	8·1	0·2	14·1	0·4
22	0 20·5	0 20·6	0 19·6	2·2	0·1	8·2	0·2	14·2	0·4
23	0 20·8	0 20·8	0 19·8	2·3	0·1	8·3	0·2	14·3	0·4
24	0 21·0	0 21·1	0 20·0	2·4	0·1	8·4	0·2	14·4	0·4
25	0 21·3	0 21·3	0 20·3	2·5	0·1	8·5	0·2	14·5	0·4
26	0 21·5	0 21·6	0 20·5	2·6	0·1	8·6	0·2	14·6	0·4
27	0 21·8	0 21·8	0 20·8	2·7	0·1	8·7	0·2	14·7	0·4
28	0 22·0	0 22·1	0 21·0	2·8	0·1	8·8	0·2	14·8	0·4
29	0 22·3	0 22·3	0 21·2	2·9	0·1	8·9	0·2	14·9	0·4
30	0 22·5	0 22·6	0 21·5	3·0	0·1	9·0	0·2	15·0	0·4
31	0 22·8	0 22·8	0 21·7	3·1	0·1	9·1	0·2	15·1	0·4
32	0 23·0	0 23·1	0 22·0	3·2	0·1	9·2	0·2	15·2	0·4
33	0 23·3	0 23·3	0 22·2	3·3	0·1	9·3	0·2	15·3	0·4
34	0 23·5	0 23·6	0 22·4	3·4	0·1	9·4	0·2	15·4	0·4
35	0 23·8	0 23·8	0 22·7	3·5	0·1	9·5	0·2	15·5	0·4
36	0 24·0	0 24·1	0 22·9	3·6	0·1	9·6	0·2	15·6	0·4
37	0 24·3	0 24·3	0 23·1	3·7	0·1	9·7	0·2	15·7	0·4
38	0 24·5	0 24·6	0 23·4	3·8	0·1	9·8	0·2	15·8	0·4
39	0 24·8	0 24·8	0 23·6	3·9	0·1	9·9	0·2	15·9	0·4
40	0 25·0	0 25·1	0 23·9	4·0	0·1	10·0	0·3	16·0	0·4
41	0 25·3	0 25·3	0 24·1	4·1	0·1	10·1	0·3	16·1	0·4
42	0 25·5	0 25·6	0 24·3	4·2	0·1	10·2	0·3	16·2	0·4
43	0 25·8	0 25·8	0 24·6	4·3	0·1	10·3	0·3	16·3	0·4
44	0 26·0	0 26·1	0 24·8	4·4	0·1	10·4	0·3	16·4	0·4
45	0 26·3	0 26·3	0 25·1	4·5	0·1	10·5	0·3	16·5	0·4
46	0 26·5	0 26·6	0 25·3	4·6	0·1	10·6	0·3	16·6	0·4
47	0 26·8	0 26·8	0 25·5	4·7	0·1	10·7	0·3	16·7	0·4
48	0 27·0	0 27·1	0 25·8	4·8	0·1	10·8	0·3	16·8	0·4
49	0 27·3	0 27·3	0 26·0	4·9	0·1	10·9	0·3	16·9	0·4
50	0 27·5	0 27·6	0 26·2	5·0	0·1	11·0	0·3	17·0	0·4
51	0 27·8	0 27·8	0 26·5	5·1	0·1	11·1	0·3	17·1	0·4
52	0 28·0	0 28·1	0 26·7	5·2	0·1	11·2	0·3	17·2	0·4
53	0 28·3	0 28·3	0 27·0	5·3	0·1	11·3	0·3	17·3	0·4
54	0 28·5	0 28·6	0 27·2	5·4	0·1	11·4	0·3	17·4	0·4
55	0 28·8	0 28·8	0 27·4	5·5	0·1	11·5	0·3	17·5	0·4
56	0 29·0	0 29·1	0 27·7	5·6	0·1	11·6	0·3	17·6	0·4
57	0 29·3	0 29·3	0 27·9	5·7	0·1	11·7	0·3	17·7	0·4
58	0 29·5	0 29·6	0 28·2	5·8	0·1	11·8	0·3	17·8	0·4
59	0 29·8	0 29·8	0 28·4	5·9	0·1	11·9	0·3	17·9	0·4
60	0 30·0	0 30·1	0 28·6	6·0	0·2	12·0	0·3	18·0	0·5

An excerpt from the Increments and Corrections pages of 'The Nautical Almanac'.

relatively easy for the purpose of computation. But the moon can be difficult not only in terms of computation, but also in sight taking, since it often appears in the sextant at a difficult angle for setting against the horizon. For beginners to celestial navigation the use of sun, stars and planets is more than sufficient for accurate navigation anywhere in the world.

Apart from the daily pages, a great deal of useful information is provided in a section at the back of *The Nautical Almanac*. In particular there are tables dealing with the Pole Star—the most useful navigational star of all. There is also a great deal of general information about heavenly bodies and the tabulations in the almanac.

The sight reduction tables

These tables come in two forms, one for marine use and one for air navigation.

The marine tables work to a greater degree of accuracy and for this reason we shall ignore the air tables in this book. Many yacht navigators will argue that on a small boat in a large ocean, such accuracy is not necessary. I refute this argument by pointing out that the time comes in every voyage where the ocean is no longer large, the coast and a landfall are looming close, and accurate navigation is essential.

The United States version of the *Sight Reduction Tables for Marine Navigation* is coded Publication 229. The equivalent British tables are coded HD605. They are, for all intents and purposes, identical. These tables are divided up into a number of volumes, each of which covers a band of latitude 15° in depth and are of use for both north and south hemispheres.

As with *The Nautical Almanac*, the *Sight Reduction Tables* are divided into two sections, the main body of the tables dealing with whole degrees, and the interpolation table inside each cover allowing for adjustments for minutes of arc. The main tables are headed by degrees of LHA at both top and bottom corners. Again, as with *The Nautical Almanac*, each section covers a double spread. The left-hand page deals with latitude and declination of the same name (north and north or south and south) and the right-hand page with latitude different (or contrary) name to declination. Thus, to find the table which you wish to use, first find the double page spread with the required LHA and then determine whether latitude has the same or different name to declination and select the left or right-hand page accordingly. Having found the page required, reading out the table is simple since it is headed by columns marked across the top with latitude and down the side with declination.

Tab values

Since the *Sight Reduction Tables* are divided into two sections for convenience, it follows that the values of LHA, Dec and Lat must be divided up before entering the tables. The main pages deal with whole degrees and these values are termed tabular values or *Tab*. Reducing them to whole degrees can be done in a number of ways and will be described in the actual computation section in Chapter 7.

Dec.	30° Hc	d	Z	31° Hc	d	Z	32° Hc	d	Z	33° Hc	d	Z	34° Hc	d	Z	35° Hc	d	Z	36° Hc	d	Z	37° Hc	d	Z	Dec.
°	° ′	′	°	° ′	′	°	° ′	′	°	° ′	′	°	° ′	′	°	° ′	′	°	° ′	′	°	° ′	′	°	°
0	36 59.0	+37.4	115.8	36 32.6	+38.3	116.4	36 05.6	+39.1	117.1	35 38.0	+40.0	117.7	35 09.8	+40.8	118.4	34 41.0	+41.6	119.0	34 11.6	+42.5	119.6	33 41.7	+43.3	120.2	0
1	37 36.4	36.8	114.8	37 10.9	37.7	115.5	36 44.7	38.7	116.2	36 18.0	39.5	116.8	35 50.6	40.4	117.5	35 22.6	41.3	118.1	34 54.1	42.1	118.7	34 25.0	42.8	119.3	1
2	38 13.2	36.3	113.8	37 48.6	37.3	114.5	37 23.4	38.2	115.2	36 57.5	39.2	115.9	36 31.0	40.0	116.6	36 03.9	40.9	117.2	35 36.2	41.7	117.8	35 07.8	42.6	118.5	2
3	38 49.5	35.7	112.8	38 25.9	36.7	113.5	38 01.6	37.7	114.2	37 36.7	38.6	114.9	37 11.0	39.6	115.6	36 44.8	40.4	116.3	36 17.9	41.3	117.0	35 50.4	42.1	117.6	3
4	39 25.2	35.2	111.7	39 02.6	36.2	112.5	38 39.3	37.2	113.2	38 15.3	38.2	114.0	37 50.6	39.1	114.7	37 25.2	40.0	115.4	36 59.2	40.9	116.1	36 32.5	41.7	116.7	4
5	40 00.4	+34.6	110.7	39 38.8	+35.7	111.5	39 16.5	+36.7	112.2	38 53.5	+37.6	113.0	38 29.7	+38.6	113.7	38 05.2	+39.6	114.4	37 40.1	+40.4	115.1	37 14.2	+41.4	115.8	5
6	40 35.0	34.0	109.6	40 14.5	35.1	110.4	39 53.2	36.1	111.2	39 31.1	37.1	112.0	39 08.3	38.1	112.7	38 44.8	39.0	113.5	38 20.5	40.0	114.2	37 55.6	40.9	114.9	6
7	41 09.0	33.4	108.5	40 49.6	34.5	109.3	40 29.3	35.5	110.2	40 08.2	36.6	110.9	39 46.4	37.6	111.7	39 23.8	38.6	112.5	39 00.5	39.5	113.2	38 36.5	40.4	114.0	7
8	41 42.4	32.8	107.4	41 24.1	33.8	108.3	41 04.8	35.0	109.1	40 44.8	36.0	109.9	40 24.0	37.0	110.7	40 02.4	38.0	111.5	39 40.0	39.0	112.3	39 16.9	40.0	113.0	8
9	42 15.2	32.0	106.3	41 57.9	33.2	107.2	41 39.8	34.3	108.0	41 20.8	35.4	108.8	41 01.0	36.5	109.7	40 40.4	37.5	110.5	40 19.0	38.5	111.3	39 56.9	39.5	112.1	9
10	42 47.2	+31.4	105.1	42 31.1	+32.5	106.0	42 14.1	+33.7	106.9	41 56.2	+34.8	107.8	41 37.5	+35.9	108.6	41 17.9	+37.0	109.4	40 57.5	+38.0	110.3	40 36.4	+38.9	111.1	10
11	43 18.6	30.6	104.0	43 03.6	31.9	104.9	42 47.8	33.0	105.8	42 31.0	34.2	106.7	42 13.4	35.2	107.5	41 54.9	36.3	108.4	41 35.5	37.4	109.2	41 15.3	38.5	110.1	11
12	43 49.2	29.9	102.8	43 35.5	31.1	103.7	43 20.8	32.3	104.6	43 05.2	33.4	105.5	42 48.6	34.6	106.4	42 31.2	35.7	107.3	42 12.9	36.8	108.2	41 53.8	37.8	109.0	12
13	44 19.1	29.1	101.6	44 06.6	30.4	102.5	43 53.1	31.6	103.5	43 38.6	32.8	104.4	43 23.2	34.0	105.3	43 06.9	35.1	106.2	42 49.7	36.2	107.1	42 31.6	37.3	108.0	13
14	44 48.2	28.3	100.4	44 37.0	29.5	101.3	44 24.7	30.8	102.3	44 11.4	32.1	103.2	43 57.2	33.3	104.2	43 42.0	34.5	105.1	43 25.9	35.6	106.0	43 08.9	36.7	106.9	14
15	45 16.5	+27.5	99.1	45 06.5	+28.8	100.1	44 55.5	+30.1	101.1	44 43.5	+31.3	102.1	44 30.5	+32.5	103.0	44 16.5	+33.7	104.0	44 01.5	+34.9	104.9	43 45.6	+36.0	105.8	15
16	45 44.0	26.7	97.8	45 35.3	28.0	98.8	45 25.6	29.2	99.9	45 14.8	30.5	100.8	45 03.0	31.8	101.8	44 50.2	33.0	102.8	44 36.4	34.2	103.8	44 21.6	35.4	104.7	16
17	46 10.7	25.7	96.5	46 03.3	27.1	97.6	45 54.8	28.5	98.6	45 45.3	29.8	99.6	45 34.8	31.0	100.6	45 23.2	32.3	101.6	45 10.6	33.5	102.6	44 57.0	34.7	103.6	17
18	46 36.4	24.8	95.2	46 30.4	26.2	96.3	46 23.3	27.5	97.3	46 15.1	28.9	98.4	46 05.8	30.2	99.4	45 55.5	31.5	100.4	45 44.1	32.8	101.4	45 31.7	34.0	102.4	18
19	47 01.2	23.9	93.9	46 56.6	25.3	95.0	46 50.8	26.7	96.0	46 44.0	28.0	97.1	46 36.0	29.4	98.1	46 27.0	30.7	99.2	46 16.9	31.9	100.2	46 05.7	33.2	101.2	19
20	47 25.1	+23.0	92.6	47 21.9	+24.4	93.6	47 17.5	+25.8	94.7	47 12.0	+27.2	95.8	47 05.4	+28.5	96.9	46 57.7	+29.8	97.9	46 48.8	+31.2	99.0	46 38.9	+32.5	100.0	20
21	47 48.1	21.9	91.2	47 46.3	23.3	92.3	47 43.3	24.8	93.4	47 39.2	26.2	94.5	47 33.9	27.6	95.6	47 27.5	29.0	96.7	47 20.0	30.3	97.7	47 11.4	31.6	98.8	21
22	48 10.0	20.9	89.8	48 09.6	22.4	90.9	48 08.1	23.8	92.0	48 05.4	25.3	93.1	48 01.5	26.7	94.3	47 56.5	28.1	95.4	47 50.3	29.5	96.5	47 43.0	30.8	97.6	22
23	48 30.9	19.9	88.4	48 32.0	21.4	89.5	48 31.9	22.9	90.6	48 30.7	24.3	91.8	48 28.2	25.7	92.9	48 24.6	27.1	94.0	48 19.8	28.5	95.2	48 13.8	29.9	96.3	23
24	48 50.8	18.8	87.0	48 53.4	20.3	88.1	48 54.8	21.8	89.2	48 55.0	23.2	90.4	48 53.9	24.8	91.5	48 51.7	26.2	92.7	48 48.3	27.6	93.8	48 43.7	29.1	95.0	24
25	49 09.6	+17.7	85.5	49 13.7	+19.2	86.7	49 16.6	+20.7	87.8	49 18.2	+22.3	89.0	49 18.7	+23.7	90.1	49 17.9	+25.2	91.3	49 15.9	+26.7	92.5	49 12.8	+28.0	93.6	25
26	49 27.3	16.6	84.0	49 32.9	18.1	85.2	49 37.3	19.6	86.4	49 40.5	21.1	87.6	49 42.4	22.7	88.7	49 43.1	24.2	89.9	49 42.6	25.7	91.1	49 40.8	27.2	92.3	26
27	49 43.9	15.4	82.6	49 51.0	17.0	83.7	49 56.9	18.6	84.9	50 01.6	20.1	86.1	50 05.1	21.6	87.3	50 07.3	23.1	88.5	50 08.3	24.6	89.7	50 08.0	26.1	90.9	27
28	49 59.3	14.2	81.1	50 08.0	15.8	82.2	50 15.5	17.4	83.4	50 21.7	18.9	84.6	50 26.7	20.5	85.8	50 30.4	22.0	87.0	50 32.9	23.5	88.3	50 34.1	25.0	89.5	28
29	50 13.5	13.1	79.5	50 23.8	14.6	80.7	50 32.9	16.2	81.9	50 40.6	17.8	83.1	50 47.2	19.3	84.4	50 52.4	20.9	85.6	50 56.4	22.5	86.8	50 59.1	24.0	88.0	29
30	50 26.6	+11.8	78.0	50 38.4	+13.5	79.2	50 49.1	+15.0	80.4	50 58.4	+16.6	81.6	51 06.5	+18.2	82.9	51 13.3	+19.8	84.1	51 18.9	+21.3	85.3	51 23.1	+22.9	86.6	30
31	50 38.4	10.7	76.5	50 51.9	12.2	77.7	51 04.1	13.7	78.9	51 15.0	15.4	80.1	51 24.7	17.0	81.3	51 33.1	18.6	82.6	51 40.2	20.2	83.8	51 46.0	21.7	85.1	31
32	50 49.1	9.3	74.9	51 04.1	10.9	76.1	51 17.8	12.6	77.3	51 30.4	14.1	78.5	51 41.7	15.7	79.8	51 51.7	17.3	81.0	52 00.4	18.9	82.3	52 07.7	20.6	83.6	32
33	50 58.4	8.1	73.4	51 15.0	9.7	74.5	51 30.4	11.3	75.8	51 44.5	12.9	77.0	51 57.4	14.5	78.2	52 09.0	16.1	79.5	52 19.3	17.7	80.8	52 28.3	19.3	82.0	33
34	51 06.5	6.8	71.8	51 24.7	8.4	73.0	51 41.7	10.0	74.2	51 57.4	11.6	75.4	52 11.9	13.2	76.6	52 25.1	14.9	77.9	52 37.0	16.5	79.2	52 47.6	18.1	80.5	34
35	51 13.3	+5.6	70.2	51 33.1	+7.1	71.4	51 51.7	+8.7	72.6	52 09.0	+10.3	73.8	52 25.1	+11.9	75.1	52 40.0	+13.5	76.3	52 53.5	+15.2	77.6	53 05.7	+16.8	78.9	35
36	51 18.9	4.2	68.6	51 40.2	5.8	69.8	52 00.4	7.3	71.0	52 19.3	9.0	72.2	52 37.0	10.6	73.4	52 53.5	12.2	74.7	53 08.7	13.8	76.0	53 22.5	15.5	77.3	36
37	51 23.1	2.9	67.0	51 46.0	4.5	68.2	52 07.7	6.1	69.4	52 28.3	7.6	70.6	52 47.6	9.3	71.8	53 05.7	10.9	73.1	53 22.5	12.6	74.4	53 38.0	14.2	75.7	37
38	51 26.0	1.7	65.4	51 50.5	3.1	66.6	52 13.8	4.7	67.7	52 35.9	6.3	68.9	52 56.9	7.9	70.2	53 16.6	9.5	71.4	53 35.1	11.1	72.7	53 52.2	12.8	74.0	38
39	51 27.7	+0.3	63.8	51 53.6	1.8	64.9	52 18.5	3.3	66.1	52 42.2	4.9	67.3	53 04.8	6.5	68.5	53 26.1	8.1	69.8	53 46.2	9.8	71.1	54 05.0	11.5	72.4	39
40	51 28.0	−1.0	62.2	51 55.4	+0.5	63.3	52 21.8	+2.0	64.5	52 47.1	+3.6	65.7	53 11.3	+5.1	66.9	53 34.2	+6.8	68.1	53 56.0	+8.4	69.4	54 16.5	+10.0	70.7	40
41	51 27.0	2.4	60.6	51 55.9	−0.8	61.7	52 23.8	+0.7	62.8	52 50.7	2.2	64.0	53 16.4	3.7	65.2	53 41.0	5.3	66.4	54 04.4	6.9	67.7	54 26.5	8.6	69.0	41
42	51 24.6	3.6	59.0	51 55.1	2.2	60.1	52 24.5	−0.7	61.2	52 52.9	+0.7	62.4	53 20.1	2.4	63.5	53 46.3	3.9	64.8	54 11.3	5.5	66.0	54 35.1	7.1	67.3	42
43	51 21.0	4.9	57.4	51 52.9	3.6	58.5	52 23.8	2.1	59.6	52 53.6	−0.5	60.7	53 22.5	+0.9	61.9	53 50.2	2.5	63.1	54 16.8	4.1	64.3	54 42.2	5.6	65.6	43
44	51 16.1	6.3	55.8	51 49.3	4.8	56.8	52 21.7	3.4	57.9	52 53.1	2.0	59.0	53 23.4	−0.5	60.2	53 52.7	+1.0	61.4	54 20.9	2.6	62.6	54 47.9	4.2	63.9	44
45	51 09.8	−7.5	54.2	51 44.5	−6.2	55.2	52 18.3	−4.8	56.3	52 51.1	−3.4	57.4	53 22.9	−1.9	58.5	53 53.7	−0.4	59.7	54 23.5	+1.1	60.9	54 52.1	+2.7	62.1	45
46	51 02.3	8.8	52.6	51 38.3	7.4	53.6	52 13.5	6.1	54.7	52 47.7	4.7	55.7	53 21.0	3.3	56.8	53 53.3	1.8	58.0	54 24.6	−0.3	59.2	54 54.8	+1.3	60.4	46
47	50 53.5	10.0	51.1	51 30.9	8.8	52.0	52 07.4	7.5	53.0	52 43.0	6.1	54.1	53 17.7	4.7	55.2	53 51.5	3.2	56.3	54 24.3	1.8	57.4	54 56.1	−0.3	58.6	47
48	50 43.5	11.3	49.5	51 22.1	10.1	50.4	51 59.9	8.8	51.4	52 36.9	7.4	52.4	53 13.0	6.0	53.5	53 48.3	4.7	54.6	54 22.5	3.2	55.7	54 55.8	1.7	56.9	48
49	50 32.2	12.5	47.9	51 12.0	11.3	48.9	51 51.1	10.0	49.8	52 29.5	8.8	50.8	53 07.0	7.5	51.8	53 43.6	6.1	52.9	54 19.3	4.7	54.0	54 54.1	3.3	55.2	49
50	50 19.7	−13.7	46.4	51 00.7	−12.5	47.3	51 41.1	−11.4	48.2	52 20.7	−10.1	49.2	52 59.5	−8.8	50.2	53 37.5	−7.5	51.2	54 14.6	−6.1	52.3	54 50.8	−4.7	53.4	50
51	50 06.0	14.9	44.9	50 48.2	13.8	45.8	51 29.7	12.6	46.6	52 10.6	11.4	47.6	52 50.7	10.2	48.5	53 30.0	8.9	49.6	54 08.5	7.6	50.6	54 46.1	6.2	51.7	51
52	49 51.1	16.0	43.4	50 34.4	14.9	44.2	51 17.1	13.8	45.1	51 59.2	12.7	46.0	52 40.5	11.5	46.9	53 21.1	10.3	47.9	54 00.9	8.9	48.9	54 39.9	7.6	50.0	52
53	49 35.1	17.2	41.9	50 19.5	16.2	42.7	51 03.3	15.1	43.5	51 46.5	14.0	44.4	52 29.0	12.8	45.3	53 10.8	11.6	46.3	53 52.0	10.4	47.2	54 32.3	9.1	48.3	53
54	49 17.9	18.3	40.4	50 03.3	17.3	41.2	50 48.2	16.3	42.0	51 32.5	15.2	42.8	52 16.2	14.1	43.7	52 59.2	12.9	44.6	53 41.6	11.8	45.6	54 23.2	10.5	46.6	54
55	48 59.6	−19.4	39.0	49 46.0	−18.4	39.7	50 31.9	−17.4	40.5	51 17.3	−16.4	41.3	52 02.1	−15.4	42.1	52 46.3	−14.3	43.0	53 29.8	−13.1	43.9	54 12.7	−11.9	44.9	55
56	48 40.2	20.4	37.5	49 27.6	19.5	38.2	50 14.5	18.6	39.0	51 00.9	17.6	39.7	51 46.7	16.6	40.6	52 32.0	15.5	41.4	53 16.7	14.4	42.3	54 00.8	13.3	43.2	56
57	48 19.8	21.5	36.1	49 08.1	20.6	36.8	49 55.9	19.7	37.5	50 43.3	18.8	38.2	51 30.1	17.7	39.0	52 16.5	16.8	39.8	53 02.3	15.7	40.7	53 47.5	14.6	41.5	57
58	47 58.3	22.4	34.7	48 47.5	21.7	35.4	49 36.2	20.8	36.0	50 24.5	19.9	36.7	51 12.4	19.0	37.5	51 59.7	18.0	38.3	52 46.6	17.0	39.1	53 32.9	16.0	39.9	58
59	47 35.9	23.5	33.3	48 25.8	22.7	33.9	49 15.4	21.9	34.6	50 04.6	21.0	35.3	50 53.4	20.2	36.0	51 41.7	19.2	36.7	52 29.6	18.3	37.5	53 16.9	17.2	38.3	59
60	47 12.4	−24.4	32.0	48 03.1	−23.6	32.6	48 53.5	−22.9	33.2	49 43.6	−22.1	33.8	50 33.2	−21.2	34.5	51 22.5	−20.4	35.2	52 11.3	−19.5	35.9	52 59.7	−18.6	36.7	60
61	46 48.0	25.3	30.6	47 39.5	24.6	31.2	48 30.6	23.8	31.8	49 21.5	23.1	32.4	50 12.0	22.4	33.0	51 02.1	21.5	33.7	51 51.8	20.6	34.4	52 41.1	19.7	35.1	61
62	46 22.7	26.3	29.3	47 14.9	25.6	29.8	48 06.8	24.9	30.4	48 58.4	24.2	31.0	49 49.6	23.4	31.6	50 40.6	22.6	32.2	51 31.2	21.8	32.9	52 21.4	21.0	33.6	62
63	45 56.4	27.1	28.0	46 49.3	26.5	28.5	47 41.9	25.8	29.0	48 34.2	25.1	29.6	49 26.2	24.4	30.1	50 18.0	23.7	30.7	51 09.4	23.0	31.4	52 00.4	22.1	32.0	63
64	45 29.3	27.9	26.7	46 22.8	27.3	27.2	47 16.1	26.8	27.7	48 09.1	26.1	28.2	49 01.8	25.4	28.7	49 54.3	24.8	29.3	50 46.4	24.0	29.9	51 38.3	23.3	30.5	64
65	45 01.4	−28.7	25.5	45 55.5	−28.2	25.9	46 49.3	−27.6	26.4	47 43.0	−27.0	26.9	48 36.4	−26.4	27.4	49 29.5	−25.7	27.9	50 22.4	−25.0	28.5	51 15.0	−24.3	29.1	65
66	44 32.7	29.6	24.2	45 27.3	29.0	24.7	46 21.7	28.5	25.1	47 16.0	27.9	25.5	48 10.0	27.3	26.0	49 03.8	26.7	26.5	49 57.4	26.1	27.0	50 50.7	25.5	27.6	66
67	44 03.1	30.3	23.0	44 58.3	29.8	23.4	45 53.2	29.3	23.8	46 48.1	28.8	24.2	47 42.7	28.3	24.7	48 37.1	27.7	25.2	49 31.3	27.1	25.7	50 25.2	26.4	26.2	67
68	43 32.8	31.0	21.8	44 28.5	30.6	22.2	45 23.9	30.1	22.6	46 19.3	29.6	23.0	47 14.4	29.1	23.4	48 09.4	28.6	23.8	49 04.2	28.0	24.3	49 58.8	27.5	24.8	68
69	43 01.8	31.7	20.6	43 57.9	31.3	21.0	44 53.8	30.8	21.3	45 49.7	30.5	21.7	46 45.3	29.9	22.1	47 40.8	29.4	22.5	48 36.2	29.0	22.9	49 31.3	28.4	23.4	69
70	42 30.1	−32.5	19.5	43 26.6	−32.1	19.8	44 23.0	−31.7	20.1	45 19.2	−31.2	20.5	46 15.4	−30.8	20.8	47 11.4	−30.3	21.2	48 07.2	−29.8	21.6	49 02.9	−29.3	22.0	70
71	41 57.6	33.1	18.4	42 54.5	32.7	18.6	43 51.3	32.3	19.0	44 48.0	31.9	19.3	45 44.6	31.5	19.6	46 41.1	31.2	20.0	47 37.4	30.7	20.3	48 33.6	30.3	20.7	71
72	41 24.5	33.7	17.2	42 21.8	33.4	17.5	43 19.0	33.1	17.8	44 16.1	32.7	18.1	45 13.1	32.4	18.4	46 09.9	31.9	18.7	47 06.7	31.5	19.1	48 03.3	31.0	19.4	72
73	40 50.8	34.3	16.1	41 48.4	34.0	16.4	42 45.9	33.7	16.6	43 43.4	33.4	16.9	44 40.7	33.0	17.2	45 38.0	32.7	17.5	46 35.2	32.3	17.8	47 32.3	32.0	18.2	73
74	40 16.5	35.0	15.1	41 14.4	34.7	15.3	42 12.2	34.3	15.5	43 10.0	34.0	15.8	44 07.7	33.7	16.0	45 05.3	33.4	16.3	46 02.9	33.1	16.6	47 00.3	32.7	16.9	74
75	39 41.5	−35.5	14.0	40 39.7	−35.2	14.2	41 37.9	−35.0	14.4	42 36.0	−34.7	14.7	43 34.0	−34.4	14.9	44 31.9	−34.1	15.1	45 29.8	−33.8	15.4	46 27.6	−33.5	15.7	75
76	39 06.0	36.0	13.0	40 04.5	35.8	13.1	41 02.9	35.6	13.3	42 01.3	35.4	13.5	42 59.6	35.1	13.8	43 57.8	34.8	14.0	44 56.0	34.5	14.2	45 54.1	34.2	14.5	76
77	38 30.0	36.6	11.9	39 28.7	36.4	12.1	40 27.3	36.1	12.3	41 25.9	35.9	12.5	42 24.5	35.7	12.7	43 23.0	35.4	12.9	44 21.5	35.2	13.1	45 19.9	34.9	13.3	77
78	37 53.4	37.1	10.9	38 52.3	36.9	11.1	39 51.2	36.7	11.2	40 50.0	36.5	11.4	41 48.8	36.3	11.6	42 47.6	36.1	11.8	43 46.3	35.8	12.0	44 45.0	35.6	12.2	78
79	37 16.3	37.5	9.9	38 15.4	37.4	10.1	39 14.5	37.2	10.2	40 13.5	37.0	10.4	41 12.5	36.8	10.5	42 11.5	36.7	10.7	43 10.5	36.5	10.8	44 09.4	36.3	11.0	79
80	36 38.8	−38.1	9.0	37 38.0	−37.9	9.1	38 37.3	−37.8	9.2	39 36.5	−37.6	9.3	40 35.7	−37.4	9.5	41 34.8	−37.2	9.6	42 34.0	−37.1	9.8	43 33.1	−36.9	9.9	80
81	36 00.7	38.5	8.0	37 00.1	38.3	8.1	37 59.5	38.2	8.2	38 58.9	38.1	8.3	39 58.3	38.0	8.4	40 57.6	37.8	8.6	41 56.9	37.6	8.7	42 56.2	37.5	8.8	81
82	35 22.2	38.9	7.1	36 21.8	38.9	7.1	37 21.3	38.7	7.2	38 20.8	38.6	7.3	39 20.3	38.4	7.4	40 19.8	38.3	7.5	41 19.3	38.2	7.7	42 18.7	38.0	7.8	82
83	34 43.3	39.4	6.1	35 42.9	39.2	6.2	36 42.6	39.2	6.3	37 42.2	39.0	6.4	38 41.9	39.0	6.4	39 41.5	38.9	6.5	40 41.1	38.8	6.6	41 40.7	38.7	6.7	83
84	34 03.9	39.7	5.2	35 03.7	39.7	5.3	36 03.4	39.5	5.3	37 03.2	39.5	5.4	38 02.9	39.4	5.5	39 02.6	39.3	5.6	40 02.3	39.2	5.6	41 02.0	39.1	5.7	84
85	33 24.2	−40.1	4.3	34 24.0	−40.0	4.4	35 23.9	−40.0	4.4	36 23.7	−40.0	4.5	37 23.5	−39.9	4.5	38 23.3	−39.8	4.6	39 23.1	−39.7	4.7	40 22.9	−39.6	4.7	85
86	32 44.1	40.5	3.4	33 44.0	40.5	3.5	34 43.9	40.5	3.5	35 43.7	40.3	3.5	36 43.6	40.3	3.6	37 43.5	40.2	3.6	38 43.4	40.2	3.7	39 43.3	40.2	3.7	86
87	32 03.6	40.9	2.5	33 03.5	40.8	2.6	34 03.4	40.7	2.6	35 03.4	40.8	2.6	36 03.3	40.7	2.7	37 03.3	40.7	2.7	38 03.2	40.7	2.7	39 03.1	40.6	2.8	87
88	31 22.7	41.2	1.7	32 22.7	41.2	1.7	33 22.7	41.2	1.7	34 22.6	41.1	1.7	35 22.6	41.1	1.8	36 22.6	41.1	1.8	37 22.5	41.0	1.8	38 22.5	41.0	1.8	88
89	30 41.5	41.5	0.8	31 41.5	41.5	0.8	32 41.5	41.5	0.9	33 41.5	41.5	0.9	34 41.5	41.5	0.9	35 41.5	41.5	0.9	36 41.5	41.5	0.9	37 41.5	41.5	0.9	89
90	30 00.0	−41.8	0.0	31 00.0	−41.8	0.0	32 00.0	−41.8	0.0	33 00.0	−41.9	0.0	34 00.0	−41.9	0.0	35 00.0	−41.9	0.0	36 00.0	−41.9	0.0	37 00.0	−41.9	0.0	90
	30°			31°			32°			33°			34°			35°			36°			37°			

46°, 314° L.H.A. LATITUDE SAME NAME AS DECLINATION

An excerpt from the main pages of the *Sight Reduction Tables for Marine Navigation*.

LATITUDE CONTRARY NAME TO DECLINATION — L.H.A. 46°, 314°

Dec.	30°			31°			32°			33°			34°			35°			36°			37°			Dec.
	Hc	d	Z	Hc	d	Z	Hc	d	Z	Hc	d	Z	Hc	d	Z	Hc	d	Z	Hc	d	Z	Hc	d	Z	
°	° ′	′	°	° ′	′	°	° ′	′	°	° ′	′	°	° ′	′	°	° ′	′	°	° ′	′	°	° ′	′	°	°
0	36 59.0	−37.8	115.8	36 32.6	−38.7	116.4	36 05.6	−39.6	117.1	35 38.0	−40.4	117.7	35 09.8	−41.3	118.4	34 41.0	−42.1	119.0	34 11.6	−42.8	119.6	33 41.7	−43.5	120.2	0
1	36 21.2	38.2	116.7	35 53.9	39.1	117.4	35 26.0	39.9	118.0	34 57.6	40.8	118.6	34 28.5	41.6	119.3	33 58.9	42.4	119.8	33 28.8	43.1	120.4	32 58.2	43.9	121.0	1
2	35 43.0	38.7	117.7	35 14.8	39.5	118.3	34 46.1	40.4	118.9	34 16.8	41.2	119.5	33 46.9	42.0	120.1	33 16.5	42.7	120.7	32 45.7	43.5	121.3	32 14.3	44.2	121.8	2
3	35 04.3	39.1	118.6	34 35.3	40.0	119.2	34 05.7	40.8	119.8	33 35.6	41.6	120.4	33 04.9	42.3	121.0	32 33.8	43.1	121.5	32 02.2	43.8	122.1	31 30.1	44.5	122.6	3
4	34 25.2	39.6	119.6	33 55.3	40.3	120.1	33 24.9	41.1	120.7	32 54.0	41.9	121.3	32 22.6	42.6	121.8	31 50.7	43.3	122.4	31 18.4	44.1	122.9	30 45.6	44.7	123.4	4
5	33 45.6	−39.9	120.5	33 15.0	−40.8	121.0	32 43.8	−41.5	121.6	32 12.1	−42.2	122.1	31 40.0	−43.0	122.7	31 07.4	−43.7	123.2	30 34.3	−44.3	123.7	30 00.9	−45.1	124.1	5
6	33 05.7	40.3	121.4	32 34.2	41.0	121.9	32 02.3	41.8	122.4	31 29.9	42.6	123.0	30 57.0	43.3	123.5	30 23.7	44.0	124.0	29 50.0	44.7	124.4	29 15.8	45.2	124.9	6
7	32 25.4	40.7	122.2	31 53.2	41.5	122.8	31 20.5	42.2	123.3	30 47.3	42.8	123.8	30 13.7	43.5	124.3	29 39.7	44.2	124.7	29 05.3	44.8	125.2	28 30.6	45.6	125.7	7
8	31 44.7	41.0	123.1	31 11.7	41.7	123.6	30 38.3	42.5	124.1	30 04.5	43.2	124.6	29 30.2	43.9	125.1	28 55.5	44.5	125.5	28 20.5	45.2	126.0	27 45.0	45.7	126.4	8
9	31 03.7	41.3	124.0	30 30.0	42.1	124.5	29 55.8	42.7	124.9	29 21.3	43.5	125.4	28 46.3	44.1	125.8	28 11.0	44.7	126.3	27 35.3	45.3	126.7	26 59.3	46.0	127.1	9
10	30 22.4	−41.7	124.8	29 47.9	−42.3	125.3	29 13.1	−43.1	125.7	28 37.8	−43.7	126.2	28 02.2	−44.3	126.6	27 26.3	−45.0	127.0	26 50.0	−45.6	127.4	26 13.3	−46.2	127.8	10
11	29 40.7	42.0	125.6	29 05.6	42.7	126.1	28 30.0	43.3	126.5	27 54.1	43.9	127.0	27 17.9	44.6	127.4	26 41.3	45.2	127.8	26 04.4	45.9	128.2	25 27.1	46.4	128.6	11
12	28 58.7	42.2	126.5	28 22.9	42.9	126.9	27 46.7	43.6	127.3	27 10.2	44.3	127.7	26 33.3	44.9	128.1	25 56.1	45.5	128.5	25 18.5	46.0	128.9	24 40.7	46.6	129.3	12
13	28 16.5	42.6	127.3	27 40.0	43.2	127.7	27 03.1	43.8	128.1	26 25.9	44.4	128.5	25 48.4	45.0	128.9	25 10.6	45.6	129.2	24 32.5	46.2	129.6	23 54.1	46.7	129.9	13
14	27 33.9	42.8	128.1	26 56.8	43.5	128.5	26 19.3	44.1	128.9	25 41.5	44.7	129.2	25 03.4	45.3	129.6	24 25.0	45.8	130.0	23 46.3	46.4	130.3	23 07.4	47.0	130.6	14
15	26 51.1	−43.1	128.8	26 13.3	−43.7	129.2	25 35.2	−44.3	129.6	24 56.8	−44.9	130.0	24 18.1	−45.5	130.3	23 39.2	−46.1	130.7	22 59.9	−46.6	131.0	22 20.4	−47.1	131.3	15
16	26 08.0	43.3	129.6	25 29.6	43.9	130.0	24 50.9	44.5	130.4	24 11.9	45.1	130.7	23 32.6	45.6	131.0	22 53.1	46.2	131.4	22 13.3	46.7	131.7	21 33.3	47.2	132.0	16
17	25 24.7	43.6	130.4	24 45.7	44.2	130.8	24 06.4	44.8	131.1	23 26.8	45.3	131.4	22 47.0	45.9	131.7	22 06.9	46.4	132.1	21 26.6	46.9	132.3	20 46.1	47.4	132.6	17
18	24 41.1	43.8	131.2	24 01.5	44.4	131.5	23 21.6	44.9	131.8	22 41.5	45.5	132.1	22 01.1	46.0	132.4	21 20.5	46.5	132.7	20 39.7	47.1	133.0	19 58.7	47.6	133.3	18
19	23 57.3	44.0	131.9	23 17.1	44.5	132.2	22 36.7	45.1	132.5	21 56.0	45.6	132.8	21 15.1	46.2	133.1	20 34.0	46.7	133.4	19 52.6	47.2	133.7	19 11.1	47.7	133.9	19
20	23 13.3	−44.2	132.6	22 32.6	−44.8	133.0	21 51.6	−45.3	133.3	21 10.4	−45.9	133.5	20 28.9	−46.3	133.8	19 47.3	−46.9	134.1	19 05.4	−47.3	134.3	18 23.4	−47.8	134.6	20
21	22 29.1	44.4	133.4	21 47.8	44.9	133.7	21 06.3	45.5	134.0	20 24.5	46.0	134.2	19 42.6	46.5	134.5	19 00.4	47.0	134.7	18 18.1	47.5	135.0	17 35.6	47.9	135.2	21
22	21 44.7	44.6	134.1	21 02.9	45.2	134.4	20 20.8	45.7	134.7	19 38.5	46.1	134.9	18 56.1	46.7	135.2	18 13.4	47.1	135.4	17 30.6	47.6	135.6	16 47.7	48.1	135.8	22
23	21 00.1	44.7	134.8	20 17.7	45.3	135.1	19 35.1	45.8	135.3	18 52.4	46.3	135.6	18 09.4	46.8	135.8	17 26.3	47.2	136.0	16 43.0	47.7	136.3	15 59.6	48.1	136.5	23
24	20 15.4	45.0	135.5	19 32.4	45.4	135.8	18 49.3	45.9	136.0	18 06.1	46.5	136.3	17 22.6	46.9	136.5	16 39.1	47.4	136.7	15 55.3	47.8	136.9	15 11.5	48.3	137.1	24
25	19 30.4	−45.1	136.2	18 47.0	−45.6	136.5	18 03.4	−46.1	136.7	17 19.6	−46.5	136.9	16 35.7	−47.0	137.1	15 51.7	−47.5	137.3	15 07.5	−47.9	137.5	14 23.2	−48.4	137.7	25
26	18 45.3	45.3	136.9	18 01.4	45.8	137.2	17 17.3	46.2	137.4	16 33.1	46.7	137.6	15 48.7	47.1	137.8	15 04.2	47.6	138.0	14 19.6	48.0	138.1	13 34.8	48.4	138.3	26
27	18 00.0	45.4	137.6	17 15.6	45.9	137.8	16 31.1	46.4	138.0	15 46.4	46.8	138.2	15 01.6	47.3	138.4	14 16.6	47.7	138.6	13 31.6	48.1	138.8	12 46.4	48.5	138.9	27
28	17 14.6	45.5	138.3	16 29.7	46.0	138.5	15 44.7	46.4	138.7	14 59.6	46.9	138.9	14 14.3	47.3	139.1	13 28.9	47.7	139.2	12 43.5	48.2	139.4	11 57.9	48.6	139.5	28
29	16 29.1	45.7	139.0	15 43.7	46.1	139.2	14 58.3	46.6	139.4	14 12.7	47.0	139.5	13 27.0	47.5	139.7	12 41.2	47.9	139.8	11 55.3	48.3	140.0	11 09.3	48.7	140.1	29
30	15 43.4	−45.8	139.7	14 57.6	−46.2	139.8	14 11.7	−46.7	140.0	13 25.7	−47.2	140.2	12 39.5	−47.5	140.3	11 53.3	−47.9	140.5	11 07.0	−48.4	140.6	10 20.6	−48.7	140.7	30
31	14 57.6	45.9	140.3	14 11.4	46.4	140.5	13 25.0	46.8	140.7	12 38.5	47.2	140.8	11 52.0	47.6	140.9	11 05.4	48.0	141.1	10 18.6	48.4	141.2	9 31.9	48.8	141.3	31
32	14 11.7	46.0	141.0	13 25.0	46.5	141.2	12 38.2	46.9	141.3	11 51.3	47.2	141.4	11 04.4	47.7	141.6	10 17.4	48.1	141.7	9 30.2	48.4	141.8	8 43.1	48.9	141.9	32
33	13 25.7	46.2	141.7	12 38.5	46.5	141.8	11 51.3	46.9	141.9	11 04.1	47.4	142.1	10 16.7	47.8	142.2	9 29.3	48.2	142.3	8 41.8	48.6	142.4	7 54.2	48.9	142.5	33
34	12 39.5	46.2	142.3	11 52.0	46.6	142.5	11 04.4	47.0	142.6	10 16.7	47.4	142.7	9 28.9	47.8	142.8	8 41.1	48.2	142.9	7 53.2	48.5	143.0	7 05.3	48.9	143.1	34
35	11 53.3	−46.3	143.0	11 05.4	−46.8	143.1	10 17.4	−47.2	143.2	9 29.3	−47.5	143.3	8 41.1	−47.9	143.4	7 52.9	−48.2	143.5	7 04.7	−48.7	143.6	6 16.4	−49.0	143.6	35
36	11 07.0	46.4	143.6	10 18.6	46.7	143.7	9 30.2	47.1	143.8	8 41.8	47.6	143.9	7 53.2	47.9	144.0	7 04.7	48.3	144.1	6 16.0	48.6	144.2	5 27.4	49.0	144.2	36
37	10 20.6	46.5	144.3	9 31.9	46.9	144.4	8 43.1	47.3	144.5	7 54.2	47.6	144.5	7 05.3	48.0	144.6	6 16.4	48.4	144.7	5 27.4	48.7	144.8	4 38.4	49.1	144.8	37
38	9 34.1	46.5	144.9	8 45.0	46.9	145.0	7 55.8	47.3	145.1	7 06.6	47.7	145.2	6 17.3	48.0	145.2	5 28.0	48.4	145.3	4 38.7	48.7	145.3	3 49.3	49.0	145.4	38
39	8 47.6	46.6	145.6	7 58.1	47.0	145.6	7 08.5	47.3	145.7	6 18.9	47.7	145.8	5 29.3	48.0	145.8	4 39.6	48.4	145.9	3 50.0	48.8	145.9	3 00.3	49.1	146.0	39
40	8 01.0	−46.7	146.2	7 11.1	−47.0	146.3	6 21.2	−47.4	146.3	5 31.2	−47.7	146.4	4 41.3	−48.1	146.4	3 51.2	−48.4	146.5	3 01.2	−48.8	146.5	2 11.2	−49.1	146.5	40
41	7 14.3	46.7	146.8	6 24.1	47.1	146.9	5 33.8	47.4	146.9	4 43.5	47.8	147.0	3 53.2	48.2	147.0	3 02.8	48.4	147.1	2 12.4	48.7	147.1	1 22.1	49.1	147.1	41
42	6 27.6	46.8	147.5	5 37.0	47.1	147.5	4 46.4	47.5	147.6	3 55.7	47.8	147.6	3 05.0	48.1	147.6	2 14.4	48.5	147.7	1 23.7	48.8	147.7	0 33.0	−49.1	147.7	42
43	5 40.8	46.8	148.1	4 49.9	47.2	148.1	3 58.9	47.5	148.2	3 07.9	47.8	148.2	2 16.9	48.1	148.2	1 25.9	48.5	148.2	0 34.9	−48.8	148.3	0 16.1	+49.2	31.7	43
44	4 54.0	46.8	148.7	4 02.7	47.2	148.8	3 11.4	47.5	148.8	2 20.1	47.8	148.8	1 28.8	48.2	148.8	0 37.4	−48.4	148.8	0 13.9	+48.8	31.2	1 05.3	49.1	31.2	44
45	4 07.2	−46.9	149.3	3 15.5	−47.2	149.4	2 23.9	−47.5	149.4	1 32.3	−47.9	149.4	0 40.6	−48.2	149.4	0 11.0	+48.5	30.6	1 02.7	+48.8	30.6	1 54.4	+49.0	30.6	45
46	3 20.3	46.9	150.0	2 28.3	47.2	150.0	1 36.4	47.6	150.0	0 44.4	−47.8	150.0	0 07.6	+48.1	30.0	0 59.5	48.5	30.0	1 51.5	48.8	30.0	2 43.4	49.1	30.0	46
47	2 33.4	46.9	150.6	1 41.1	47.2	150.6	0 48.8	47.5	150.6	0 03.4	+47.9	29.4	0 55.7	48.2	29.4	1 48.0	48.5	29.4	2 40.3	48.7	29.4	3 32.5	49.1	29.4	47
48	1 46.5	46.9	151.2	0 53.9	47.2	151.2	0 01.3	−47.5	151.2	0 51.3	47.8	28.8	1 43.9	48.1	28.8	2 36.5	48.4	28.8	3 29.0	48.8	28.8	4 21.6	49.0	28.9	48
49	0 59.6	47.0	151.8	0 06.7	−47.3	151.8	0 46.2	+47.6	28.2	1 39.1	47.9	28.2	2 32.0	48.1	28.2	3 24.9	48.4	28.2	4 17.8	48.7	28.2	5 10.6	49.0	28.3	49
50	0 12.6	−46.9	152.5	0 40.6	+47.2	27.5	1 33.8	+47.5	27.6	2 27.0	+47.8	27.6	3 20.1	+48.1	27.6	4 13.3	+48.4	27.6	5 06.5	+48.6	27.7	5 59.6	+49.0	27.7	50
51	0 34.3	+46.9	26.9	1 27.8	47.2	26.9	2 21.3	47.5	26.9	3 14.8	47.8	27.0	4 08.2	48.1	27.0	5 01.7	48.4	27.0	5 55.1	48.7	27.1	6 48.6	48.9	27.1	51
52	1 21.2	47.0	26.3	2 15.0	47.2	26.3	3 08.8	47.5	26.3	4 02.6	47.7	26.4	4 56.3	48.1	26.4	5 50.1	48.3	26.4	6 43.8	48.6	26.5	7 37.5	48.8	26.5	52
53	2 08.2	46.9	25.7	3 02.2	47.2	25.7	3 56.3	47.5	25.7	4 50.3	47.8	25.8	5 44.4	48.0	25.8	6 38.4	48.3	25.8	7 32.4	48.5	25.9	8 26.3	48.9	26.0	53
54	2 55.1	46.8	25.0	3 49.4	47.2	25.1	4 43.8	47.4	25.1	5 38.1	47.7	25.1	6 32.4	47.9	25.2	7 26.7	48.2	25.2	8 20.9	48.5	25.3	9 15.2	48.7	25.4	54
55	3 41.9	+46.9	24.4	4 36.6	+47.1	24.5	5 31.2	+47.4	24.5	6 25.8	+47.6	24.5	7 20.3	+48.0	24.6	8 14.9	+48.2	24.6	9 09.4	+48.5	24.7	10 03.9	+48.7	24.8	55
56	4 28.8	46.8	23.8	5 23.7	47.1	23.8	6 18.6	47.3	23.9	7 13.4	47.6	23.9	8 08.3	47.8	24.0	9 03.1	48.1	24.0	9 57.9	48.3	24.1	10 52.6	48.6	24.2	56
57	5 15.6	46.8	23.2	6 10.8	47.0	23.2	7 05.9	47.3	23.3	8 01.0	47.6	23.3	8 56.1	47.8	23.4	9 51.2	48.0	23.4	10 46.2	48.3	23.5	11 41.2	48.6	23.6	57
58	6 02.4	46.8	22.5	6 57.8	47.0	22.6	7 53.2	47.3	22.6	8 48.6	47.5	22.7	9 43.9	47.8	22.8	10 39.2	48.0	22.8	11 34.5	48.3	22.9	12 29.8	48.5	23.0	58
59	6 49.2	46.7	21.9	7 44.8	47.0	22.0	8 40.5	47.1	22.0	9 36.1	47.4	22.1	10 31.7	47.6	22.1	11 27.2	47.9	22.2	12 22.8	48.1	22.3	13 18.3	48.3	22.4	59
60	7 35.9	+46.6	21.3	8 31.8	+46.8	21.3	9 27.6	+47.2	21.4	10 23.5	+47.4	21.4	11 19.3	+47.6	21.5	12 15.1	+47.9	21.6	13 10.9	+48.1	21.7	14 06.6	+48.3	21.8	60
61	8 22.5	46.6	20.6	9 18.6	46.9	20.7	10 14.8	47.0	20.8	11 10.9	47.2	20.8	12 06.9	47.5	20.9	13 03.0	47.7	21.0	13 59.0	47.9	21.1	14 54.9	48.2	21.2	61
62	9 09.1	46.5	20.0	10 05.5	46.7	20.1	11 01.8	47.0	20.1	11 58.1	47.2	20.2	12 54.4	47.4	20.3	13 50.7	47.6	20.4	14 46.9	47.9	20.4	15 43.1	48.1	20.5	62
63	9 55.6	46.4	19.4	10 52.2	46.7	19.4	11 48.8	46.8	19.5	12 45.3	47.1	19.6	13 41.8	47.4	19.6	14 38.3	47.6	19.7	15 34.8	47.8	19.8	16 31.2	48.0	19.9	63
64	10 42.0	46.4	18.7	11 38.9	46.5	18.8	12 35.6	46.8	18.9	13 32.4	47.0	18.9	14 29.2	47.2	19.0	15 25.9	47.4	19.1	16 22.6	47.6	19.2	17 19.2	47.9	19.3	64
65	11 28.4	+46.3	18.1	12 25.4	+46.5	18.1	13 22.4	+46.7	18.2	14 19.4	+46.9	18.3	15 16.4	+47.1	18.4	16 13.3	+47.3	18.5	17 10.2	+47.5	18.6	18 07.1	+47.7	18.7	65
66	12 14.7	46.1	17.4	13 11.9	46.4	17.5	14 09.1	46.6	17.6	15 06.3	46.8	17.6	16 03.5	47.0	17.7	17 00.6	47.2	17.8	17 57.7	47.4	17.9	18 54.8	47.6	18.0	66
67	13 00.8	46.1	16.8	13 58.3	46.3	16.8	14 55.7	46.5	16.9	15 53.1	46.7	17.0	16 50.5	46.8	17.1	17 47.8	47.1	17.2	18 45.1	47.3	17.3	19 42.4	47.5	17.4	67
68	13 46.9	46.0	16.1	14 44.6	46.1	16.2	15 42.2	46.3	16.3	16 39.8	46.5	16.3	17 37.3	46.8	16.4	18 34.9	46.9	16.5	19 32.4	47.1	16.6	20 29.9	47.3	16.7	68
69	14 32.9	45.9	15.4	15 30.7	46.1	15.5	16 28.5	46.3	15.6	17 26.3	46.4	15.7	18 24.1	46.6	15.8	19 21.8	46.8	15.9	20 19.5	47.0	16.0	21 17.2	47.1	16.1	69
70	15 18.8	+45.7	14.8	16 16.8	+45.9	14.9	17 14.8	+46.1	14.9	18 12.7	+46.3	15.0	19 10.7	+46.4	15.1	20 08.6	+46.6	15.2	21 06.5	+46.8	15.3	22 04.3	+47.0	15.4	70
71	16 04.5	45.6	14.1	17 02.7	45.8	14.2	18 00.9	45.9	14.3	18 59.0	46.1	14.3	19 57.1	46.3	14.4	20 55.2	46.5	14.5	21 53.3	46.6	14.6	22 51.3	46.9	14.7	71
72	16 50.1	45.5	13.4	17 48.5	45.6	13.5	18 46.8	45.8	13.6	19 45.1	46.0	13.7	20 43.4	46.2	13.7	21 41.7	46.3	13.8	22 39.9	46.5	13.9	23 38.2	46.6	14.0	72
73	17 35.6	45.4	12.7	18 34.1	45.5	12.8	19 32.6	45.7	12.9	20 31.1	45.8	13.0	21 29.6	45.9	13.1	22 28.0	46.1	13.2	23 26.4	46.3	13.3	24 24.8	46.5	13.4	73
74	18 21.0	45.2	12.1	19 19.6	45.4	12.1	20 18.3	45.5	12.2	21 16.9	45.7	12.3	22 15.5	45.8	12.4	23 14.1	46.0	12.5	24 12.7	46.1	12.6	25 11.3	46.2	12.7	74
75	19 06.2	+45.0	11.4	20 05.0	+45.1	11.4	21 03.8	+45.3	11.5	22 02.6	+45.4	11.6	23 01.3	+45.6	11.7	24 00.1	+45.7	11.8	24 58.8	+45.9	11.9	25 57.5	+46.1	12.0	75
76	19 51.2	44.9	10.7	20 50.1	45.0	10.7	21 49.1	45.1	10.8	22 48.0	45.3	10.9	23 46.9	45.4	11.0	24 45.8	45.6	11.0	25 44.7	45.7	11.1	26 43.6	45.8	11.2	76
77	20 36.1	44.6	10.0	21 35.1	44.9	10.0	22 34.2	45.0	10.1	23 33.3	45.1	10.2	24 32.3	45.2	10.2	25 31.4	45.3	10.3	26 30.4	45.5	10.4	27 29.4	45.6	10.5	77
78	21 20.7	44.6	9.2	22 20.0	44.6	9.3	23 19.2	44.7	9.4	24 18.4	44.8	9.4	25 17.5	45.0	9.5	26 16.7	45.1	9.6	27 15.9	45.2	9.7	28 15.0	45.4	9.8	78
79	22 05.3	44.3	8.5	23 04.6	44.4	8.6	24 03.9	44.6	8.6	25 03.2	44.7	8.7	26 02.5	44.8	8.8	27 01.8	44.9	8.9	28 01.1	45.0	8.9	29 00.4	45.1	9.0	79
80	22 49.6	+44.1	7.8	23 49.0	+44.2	7.8	24 48.5	+44.3	7.9	25 47.9	+44.4	8.0	26 47.3	+44.5	8.0	27 46.7	+44.6	8.1	28 46.1	+44.7	8.2	29 45.5	+44.8	8.3	80
81	23 33.7	43.9	7.1	24 33.2	44.0	7.1	25 32.8	44.1	7.2	26 32.3	44.2	7.2	27 31.8	44.3	7.3	28 31.3	44.4	7.4	29 30.8	44.5	7.4	30 30.3	44.6	7.5	81
82	24 17.6	43.7	6.3	25 17.2	43.8	6.4	26 16.9	43.8	6.4	27 16.5	43.9	6.5	28 16.1	44.0	6.5	29 15.7	44.1	6.6	30 15.3	44.2	6.7	31 14.9	44.3	6.7	82
83	25 01.3	43.4	5.6	26 01.0	43.5	5.6	27 00.7	43.6	5.6	28 00.4	43.7	5.7	29 00.1	43.8	5.8	29 59.8	43.8	5.8	30 59.5	43.9	5.9	31 59.2	44.0	5.9	83
84	25 44.7	43.3	4.8	26 44.5	43.3	4.8	27 44.3	43.4	4.9	28 44.1	43.4	4.9	29 43.9	43.4	5.0	30 43.6	43.6	5.0	31 43.4	43.6	5.1	32 43.2	43.7	5.1	84
85	26 28.0	+42.9	4.0	27 27.8	+43.0	4.1	28 27.7	+43.0	4.1	29 27.5	+43.1	4.1	30 27.3	+43.2	4.2	31 27.2	+43.2	4.2	32 27.0	+43.3	4.3	33 26.9	+43.3	4.3	85
86	27 10.9	42.7	3.2	28 10.8	42.8	3.3	29 10.7	42.8	3.3	30 10.6	42.8	3.3	31 10.5	42.9	3.4	32 10.4	42.9	3.4	33 10.3	43.0	3.4	34 10.2	43.0	3.5	86
87	27 53.6	42.4	2.4	28 53.6	42.4	2.5	29 53.5	42.5	2.5	30 53.4	42.6	2.5	31 53.4	42.5	2.5	32 53.3	42.6	2.6	33 53.3	42.6	2.6	34 53.2	42.6	2.6	87
88	28 36.0	42.2	1.6	29 36.0	42.2	1.7	30 36.0	42.2	1.7	31 36.0	42.1	1.7	32 35.9	42.2	1.7	33 35.9	42.2	1.7	34 35.9	42.2	1.7	35 35.8	42.3	1.8	88
89	29 18.2	41.8	0.8	30 18.2	41.8	0.8	31 18.2	41.8	0.8	32 18.1	41.9	0.9	33 18.1	41.9	0.9	34 18.1	41.9	0.9	35 18.1	41.9	0.9	36 18.1	41.9	0.9	89
90	30 00.0	+41.5	0.0	31 00.0	+41.5	0.0	32 00.0	+41.5	0.0	33 00.0	+41.5	0.0	34 00.0	+41.5	0.0	35 00.0	+41.5	0.0	36 00.0	+41.5	0.0	37 00.0	+41.5	0.0	90
	30°			31°			32°			33°			34°			35°			36°			37°			

S. Lat. { L.H.A. greater than 180°......Zn=180°−Z; L.H.A. less than 180°............Zn=180°+Z

LATITUDE SAME NAME AS DECLINATION — L.H.A. 134°, 226°

An excerpt from the main pages of the *Sight Reduction Tables for Marine Navigation*.

Assumed Lat and Long

In order to reduce all values to whole degrees for use in the table, the DR Lat and Long must be adjusted also. The term given to these adjusted values is *assumed* Lat and Long indicated by the use of the letter a—aLat, aLon. In the case of the latitude it is simply a question of adopting the nearest whole degree. The aLon is a little more involved and is described in detail in the computation section of Chapter 7.

Sight reduction factors

Having entered the main pages of the *Sight Reduction Table* with LHA, lat and dec, three factors are obtained which are coded Hc, d and Z.

Hc

This is the computed altitude. The true or observed altitude found by means of the sextant is coded Ho. Thus it can be said that H is the code term for altitude, the 'o' and 'c' relating to observed and calculated altitudes respectively. Since the calculated altitude is derived from a DR or estimated position and the observed altitude is obtained from the sextant, one could say broadly that the difference between Hc and Ho is the difference between where you think the boat is and where your observations show is the precise position of the boat.

'd'

The correct name for this symbol is *altitude difference*, but since it is only a stepping stone between the main pages and the interpolation tables, and is discarded after use, it is more commonly just referred to as 'd' and left at that.

Z

Z stands for the azimuth angle, which is best described as the bearing of the celestial body measured eastwards or westwards through 180° from north or south. By addition or subtraction of 180° it is converted into a compass bearing or azimuth (coded Zn) which can then be used for plotting.

Interpolation table

The Interpolation Tables on the inside front and back covers of the *Sight Reduction Tables* need a little practice in reading and are dealt with in detail in the computation section of Chapter 7.

INTERPOLATION TABLE

Dec. Inc.	Altitude Difference (d): Tens 10′ 20′ 30′ 40′ 50′	Decimals	Units 0′ 1′ 2′ 3′ 4′ 5′ 6′ 7′ 8′ 9′	Double Second Diff. and Corr.
[illegible]	[illegible]	[illegible]	[illegible]	[illegible]

The Double-Second-Difference correction (Corr.) is always to be added to the tabulated altitude.

INTERPOLATION TABLE

Dec. Inc.	Altitude Difference (d): Tens 10′ 20′ 30′ 40′ 50′	Decimals	Units 0′ 1′ 2′ 3′ 4′ 5′ 6′ 7′ 8′ 9′	Double Second Diff. and Corr.
[illegible]	[illegible]	[illegible]	[illegible]	[illegible]

The Double-Second-Difference correction (Corr.) is always to be added to the tabulated altitude.

An excerpt from the Interpolation pages of the *Sight Reduction Tables for Marine Navigation*.

Chapter 7 **The Computations**

Leaving aside the noon latitude sight (at the bottom of the form) for the moment, the Sight Reduction Form is divided into four main sections. At the top is the entry of information gathered in taking the sight, including the date, LMT, name of the star/planet/sun, GMT and the DR position. The next section deals with the correction of the sextant altitude to obtain the true altitude (Ho). The third section deals with the information taken out of *The Nautical Almanac* and reduces it to the required data for entry into the *Marine Sight Reduction Tables*. The final section, using these sight reduction tables, resolves the sight to an azimuth (Zn) and the difference between Hc and Ho which is termed the *intercept*. These are the two factors required for plotting.

Probably the best way to explain each factor of the computation is to work an example, and initially we shall use an afternoon sun sight.

Section 1

This is the information gathered prior to the calculation including the date (December 15 1984), the GMT of the sight (22h 01m 40s), DR position (Lat 30°17′S and Long 105°05′W). Other information which will be known but not entered in Section 1 is the sextant altitude (48°26′), the index error (1.0′ on the arc) and the height-of-eye above sea level (3 meters). Now the computation begins in earnest and I shall describe each factor in detail as we progress through each section.

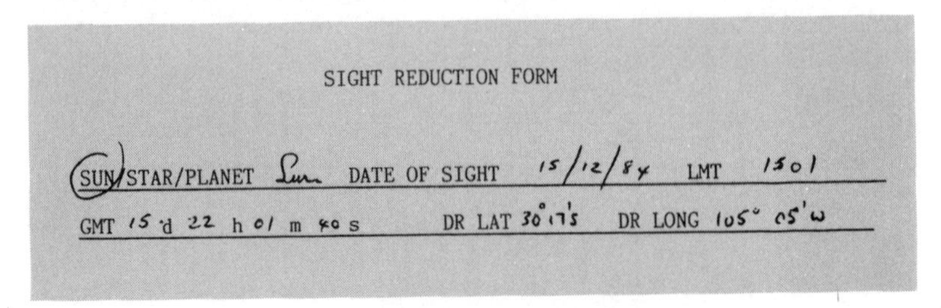

SIGHT REDUCTION FORM

SUN/STAR/PLANET Sun DATE OF SIGHT 15/12/84 LMT 1501

GMT 15 d 22 h 01 m 40 s DR LAT 30°17′S DR LONG 105° 05′W

Sextant

Sext alt	48° 26′	
I.E.	1′	+ ⊖
Obs alt	48° 25′	
Dip	3′	-
App Alt	48° 22′	
Alt Corr	15.4	+
TRUE ALT	48° 37.4′	(Ho)

Section 11

1 To the sextant alt (48°26′) apply the index error (-1.0′) to obtain the obs alt (48°25′).

2 From the inside cover of *The Nautical Almanac* obtain the dip for a height-of-eye of 3 meters (-3.0′) and apply it to the obs alt to find the app alt (48°22′).

3 From the same page in *The Nautical Almanac* take out the altitude correction for the sun in the column October to March using the app alt as closely as possible. Apply this alt corr (+15.4′) to obtain the true alt or Ho (48°37.4′).

Section 111

1 Open the main pages of *The Nautical Almanac* to December 15 in the Sun column and take out the GHA for 22 hours GMT (151°07.7′).

2 Turn to the tinted pages at the back of the *Nautical Almanac* and take out the increment for the minutes and seconds of GMT (01 minutes and 40 seconds). The minutes are located at the top of the page in bold type and the seconds in the first column. Ensure that the correction comes from the Sun column and not Aries or Moon (0°25′).

Note: In the case of a star the GHA and increment would have been taken out for Aries at this point. Then, back on the daily pages, the SHA for the star would be taken out of the column marked 'Stars'. Since this is a sun sight the SHA is not applicable.

Almanac

GHA 151° 07.7′ (Aries for stars)

Inc 25.0′ +

SHA — — + (Stars only)

Total GHA 151° 32.7′

a LON 105° 32.7′ − +

Tab LHA 46° 00.0′

TAB DEC 23° S DEC INC 18.7′ a LAT 30° S

3 Add all these readouts to find the total GHA for the GMT of 22h 01m 40s (151°32.7′).

4 To find the LHA, the GHA must be added to or subtracted from the DR Long. East longitude is added and west longitude subtracted. However, for the sight reduction tables a *Tab* (whole degrees only) LHA is required, therefore a longitude must be assumed which, while being as close as possible to the DR Long will, when applied to the GHA, eliminate the minutes and leave only whole degrees. In this case, because we are subtracting, an aLon of 105°32.7′ will do the trick. By applying this aLon we obtain a Tab LHA of 46°.

5 Return to the daily pages on 15 December and take out the declination for the GMT. A little interpolation may be required here, but since the declination changes only slightly in one hour, it can be done by eye. On the form enter the degrees of declination (23°S) against Tab Dec and the minutes against the heading Dec Inc (18.7′).

6 The aLat, as described earlier, is the DR latitude to the nearest whole degree (30°S).

Section 1V

1 Open the *Marine Sight Reduction Tables* to the double page

```
Tables (HO 229)
                         Hc 48° 30.9'    d + 19.9        Z 88.4°

corr (1)..3.1'..
corr (2)..3.1'..
                                   (+)
Tot corr 6.2' >>>>>>       6.2' (+ according to d)

                         Hc 48° 37.1'

                         Ho 48° 37.4'

        INTERCEPT        0.3' (T/A)        AZIMUTH (Zn) 268.4°
```

headed 46° LHA. Since Dec and Lat are both South, the left-hand page (latitude same name as declination) will be the one used.

2 On this page, in the column headed 30° aLat and against the Dec (side columns) of 23° take out the three factors and enter them on the Sight Form (Hc 48°30.9′, d + 19.9′, Z 88.4°).

3 While still on these pages look down the bottom margin where small print indicates the application of Z to find Zn. Since we are in South latitudes and the LHA is less than 180°, Zn 180° + Z. We can therefore complete the right-hand section by finding Zn (268.4°).

4 Now open the *Sight Reduction Tables* at the inside cover (in this case the front). This table uses the arguments Dec Inc (18.7′) and d (19.9′) found earlier. Whereas the main pages dealt with whole degrees, now the interpolation table caters for the minutes and decimals. With the Dec Inc in the left-hand column (18.7′) and the tens of 'd' (10), the first correction is read out under the tens column (3.1′).

5 Having used 10 of the 19.9′ in d, the remaining 9.9′ are found under the units and decimals section. Follow the unit 9′ column down to the block of decimals that covers the Dec Inc (18.0′ to 19.0′). Against the decimal.9 in the column 9′ will be found the second correction (3.1′). In other words we are looking for the 9.9 second correction in the 18′ block where we found the first correction.

6 Add these two corrections to find the total correction and then apply to the Hc according to the sign at 'd' (in this case +). The result is the corrected Hc (48°37.1′).

7 Bring down the Ho found in Section 11 (48°37.4′) and subtract Ho from Hc (or vice versa, as the case may be) to find the Int (0.3′).

8 This is named towards or away from the sun in order to facilitate plotting (see later). One way of remembering whether the intercept is named towards or away is to remember the first

A full sight.

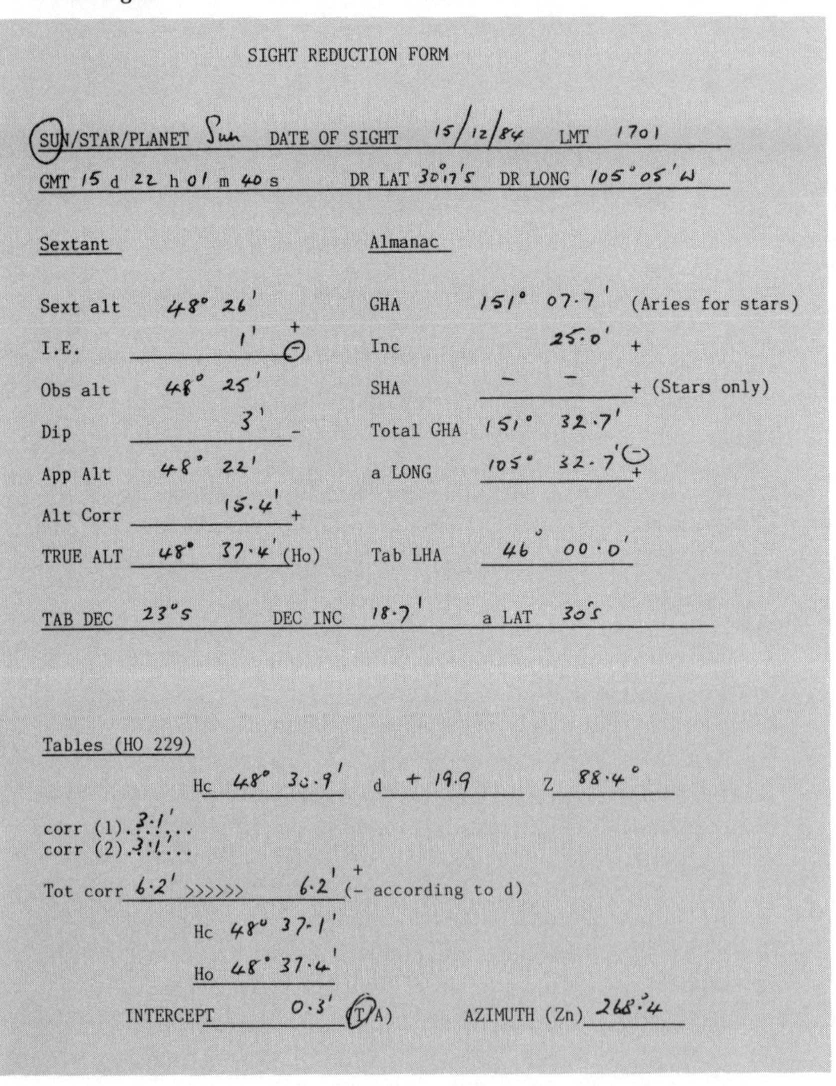

SIGHT REDUCTION FORM

SUN/STAR/PLANET Sun DATE OF SIGHT 15/12/84 LMT 1701

GMT 15 d 22 h 01 m 40 s DR LAT 30°17′S DR LONG 105°05′W

Sextant		Almanac	
Sext alt	48° 26′	GHA	151° 07.7′ (Aries for stars)
I.E.	1′ + (−)	Inc	25.0′ +
Obs alt	48° 25′	SHA	— — + (Stars only)
Dip	3′ −	Total GHA	151° 32.7′
App Alt	48° 22′	a LONG	105° 32.7′ (−) +
Alt Corr	15.4′ +		
TRUE ALT	48° 37.4′ (Ho)	Tab LHA	46° 00.0′

TAB DEC 23°S DEC INC 18.7′ a LAT 30°S

Tables (HO 229)

Hc 48° 30.9′ d + 19.9 Z 88.4°

corr (1) 3.1′

corr (2) 3.1′

Tot corr 6.2′ >>>>>> 6.2′ (+ − according to d)

Hc 48° 37.1′

Ho 48° 37.4′

INTERCEPT 0.3′ (T/A) AZIMUTH (Zn) 268.4°

three letters of the author's name (TOG—Towards if the Observed is Greatest!). Intercept 0.3′ Towards.

This method of sight reduction can be used with any heavenly body where a position line is required for plotting. This applies to all sights except the noon latitude sight. Morning and evening star sights and morning and afternoon sun sights are all resolved with the sight reduction method. The information acquired—azimuth (Zn) and intercept (int)—is all that is required from each sight to enable it to be plotted as described in Chapter 8.

The noon sight

The noon sight is the simplest of all sights. In a matter of two or three lines of calculations it offers not just a position line but a complete and accurate latitude. This in itself is not a fix position, of course, since a longitude is required as well as a latitude. However, since the morning sun sight will have been taken only a few hours previously, its position line can be run forward and crossed with the noon latitude as a running fix, which is usually sufficiently accurate to provide a full noon position of latitude and longitude.

The noon latitude sight form at the foot of the sight reduction form indicates the simplicity of this sight. The left-hand side of the form is merely the correction of the sextant altitude as is done for all sights. The noon latitude sight itself—correctly termed a *meridian passage sight*—takes up only the few lines on the right.

As described earlier, taking the noon sight is also easier than taking other sights since the sun is virtually static as it makes its

The sun appears to rise in the horizon mirror of the sextant, pause as it crosses the meridian, then fall. The meridian passage altitude is taken at the peak or zenith.

meridian passage. You should begin the sight some minutes before noon is due to allow for possible error in the calculation of the time of noon (see Chapter 5). The sun at this point will still be rising slowly towards its zenith and adjustment of the micrometer screw on the sextant will be necessary to hold it on the horizon. As the meridian passage approaches, this movement will slow considerably to the point where the sun appears to be static and the micrometer requires no adjustment to hold it on the horizon. Then, a minute or two later, it will appear to have dipped below the horizon and commenced its descent to the west. The meridian passage is over. Providing you do not touch the micrometer the sextant will have recorded its maximum altitude.

Step one, as with all sextant sights, is to correct this altitude on the left-hand side of the sight form to obtain Ho. The procedure for reducing the latitude sight on the right-hand side of the form is as follows.

1 Subtract Ho from 90° to obtain the zenith distance (ZX). ZX is named according to which side of the sun you are. That is, if the sun is to the north of you, then ZX is named South. If you are navigating in the Northern Hemisphere then most likely the sun will be to the south, which means that ZX will be named North.
2 From *The Nautical Almanac* take out the declination of the sun from the daily pages for the GMT of the noon sight. Since

A typical example of a latitude by meridian altitude (noon) sight.

NOON LATITUDE

Sext alt	77° 42·0'			90°00'	
I.E.	1·0 (−)	+	Ho	77° 54'	−
Obs alt	77° 41·0'		ZX	12° 06	S
Dip	3·0	−	Dec	23° 14·4 (+)	− S
App alt	77° 38·0'				
Alt Corr	16·0'	+	NOON LAT	35° 20·4'	S
Ho	77° 54·0'				

declination changes only slowly, accurate timing is not as essential with the noon sight as it is with other sights. The GMT should be read to the nearest minute, seconds are not necessary.

3 Apply the declination to the ZX in the following way—add if they are of the same name or subtract the smaller from the larger if they are of different names—to obtain the noon latitude of the boat.

It is not necessary to use the plotting sheet for this sight, since the latitude can be laid directly on the chart. However, if the scale of the chart is very small it may make the run up of the morning sight somewhat difficult, in which case a plotting sheet should be used and the final running fix position transferred to the ocean chart.

Chapter 8 **Plotting**

The simplest and most common method of plotting star sights is by using the Position Plotting Sheets obtainable from chart agents. These are, in effect, a 'blow up' of a section of an ocean chart with no land masses and the scales enlarged to the point where laying off bearings and azimuths, measuring intercepts and transferring position lines can be done easily and with room to move.

The plotting sheets are restricted to bands of latitude usually about 3° to 4° wide and so designed that they can be used in Northern or Southern Hemispheres simply by turning them upside-down. The longitude scale is not marked so they can be adopted for any longitude around the world. This makes the sheets very flexible and re-useable with the only limiting factor being the band of latitude. For convenience in plotting, compass roses are placed strategically across them as they are on normal charts.

The DR position

The assumed position is the point used for plotting with the sight reduction method. However, I find it convenient to also plot the DR position on the plotting sheet so that when the fix has been obtained, reference to the DR will indicate set and drift and other factors which have affected the boat's progress. This, of course, can be done on the normal ocean chart, but once again the scale is such that it makes accurate working very difficult, and the expanded scale of the plotting sheet allows better measurement of bearings and distances.

The assumed position

All azimuths and intercepts must be plotted from the assumed position when using the sight reduction method. Where three or four star sights have been taken at the one time, they will have a common aLat but not a common aLon, since the aLon was adapted to obtain a Tab LHA. Since each star would have a different GHA, it would obviously have a different aLon. It is vitally important that the azimuth is laid off from the assumed longitude for each individual star, which means that each plot will begin at a different point along the common assumed latitude.

Single position lines

The procedure for laying off a single position line from a sight reduction computation is as follows:

1 Locate on the plotting sheet the assumed latitude and longitude position.
2 With the parallel rules draw through this position the azimuth (Zn). Draw this azimuth from the assumed position towards the heavenly body and also away from it on the reciprocal bearing.

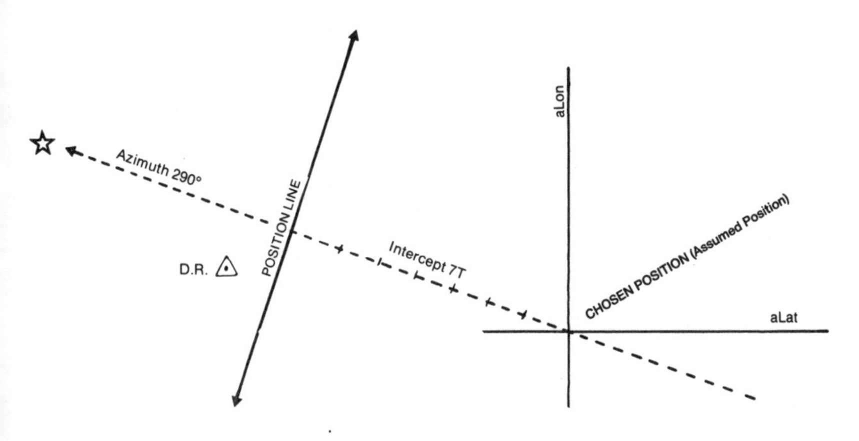

A single position line plot.

3 Using the latitude scale on the plotting sheet, measure the intercept from the assumed position towards or away from the heavenly body along the azimuth.
4 Through this point draw a line at right angles to the azimuth. This is the position line, and the boat's position must lie somewhere along it.

Full star plot

The procedure for plotting three or more stars is only a repetition of plotting a single position line. However, since there are differing assumed longitudes, it is worth running through the routine for a full three-star plot.

1 Across the plotting sheet draw the aLat.
2 Establish the aLon for the first star on this assumed latitude.
3 Through this assumed position draw the azimuth both towards and away from the star as described above.

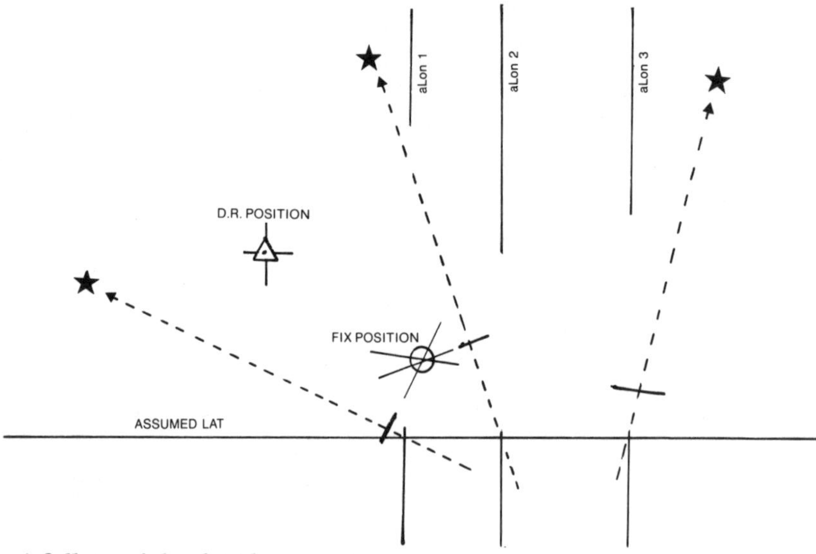

A full star sight plot (three stars).

4 Measure the intercept towards or away from the star along the azimuth.
5 Through this point draw the position line at right angles and extend it for some distance.
6 Repeat with the second star establishing its assumed position along the aLat with its aLon as before.
7 Repeat with all other stars until plotting is completed.
8 The point at which all position lines intersect or nearly intersect is the fix of the boat's position. It can be measured in latitude and longitude and transferred to the working chart.

Celestial running fix

This is most commonly used during the daytime when only single position lines are available by sun sight. However, it can be used at any time and with any heavenly body. The principle is exactly the same as the running fix in coastal navigation where two separate bearings are run together to make a fix. For sake of example, we will assume that we are running a morning sun sight position line to a noon sun latitude position line along a given course:

1 Plot the DR position on the plotting sheet for morning sight time and draw the courseline through this position extending it well in both directions.

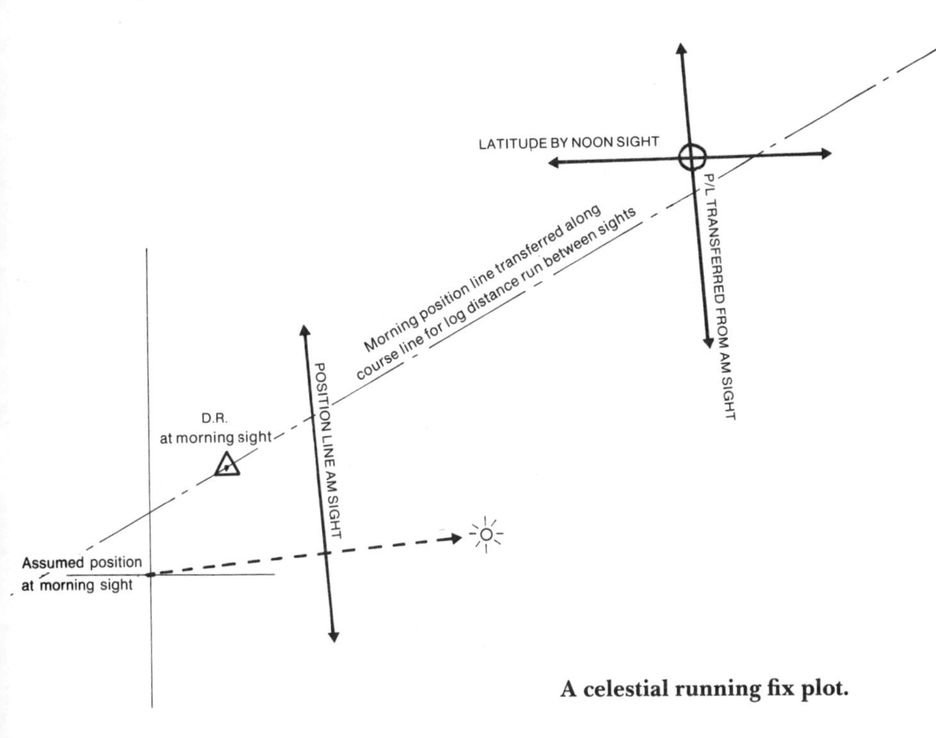

A celestial running fix plot.

2 Using the assumed latitude and longitude (not DR) plot the position line for morning sun sight as described above.

3 At noon plot the noon latitude.

4 Transfer the morning position line along the courseline from its intersection for the log distance run since that sight was taken until noon.

5 Where the transferred morning position line cuts the noon latitude is the fix of the boat's position at noon.
6 If the boat alters course between the morning sight and the noon latitude, it will be necessary to plot the different courses and distances along them and transfer the morning sun sight position line along each section.

Set and drift

As with coastal navigation, the boat is unlikely to maintain her courseline consistently, being affected by a number of factors, foremost of which will probably be the drift of the ocean current. To ascertain the amount of drift and its direction, the fix and DR positions must be plotted. Then the line from the DR position to the fix position at the same time represents the direction and distance of the drift that has been experienced since the last fix position was established.

As with coastal navigation, the difference between DR and fix positions represents the set and drift of the current experienced during the passage.

Chapter 9 **Other Celestial Work**

The 'navigator's day' described in Chapter 1 involves the use of a number of fixes commencing and finishing with full star sights at dawn and dusk and using sun position lines during the day. However, there are other forms of sight computation which can be used under certain circumstances and which often fill a need. A typical example would be the ex-meridian sight which is used to obtain a noon latitude when the sun is obscured right on noon. Depending on latitude and declination, this sight can give an accurate noon latitude sometimes up to half an hour on either side of actual noon.

The noon latitude by sun (meridian altitude) can also be applied to other bodies, providing a sight can be taken as they cross the meridian. Stars and planets are rarely in this position since they must make their meridian passage in the brief twilight period when both stars and horizon are visible. However, by comparing the boat's DR longitude and the GHA of the star or planet, you can determine just when the meridian passage of any heavenly body is made and, if it coincides with star sight taking time, a latitude by meridian passage can supplement the full star fix. I do not propose to go into the many computations that can be used for different types of sights, but merely in this chapter to mention a few useful celestial workings which may supplement those described in our routine sight-taking day.

Latitude by ex-meridian reduction

In order to compute this sight, a set of nautical tables is required. These tables provide a correction to the sextant altitude taken on either side of meridian passage. This correction converts the ex-meridian altitude to that which would have been obtained had the sight been taken right on meridian passage.

The ex-meridian sight can only be used within certain limits,

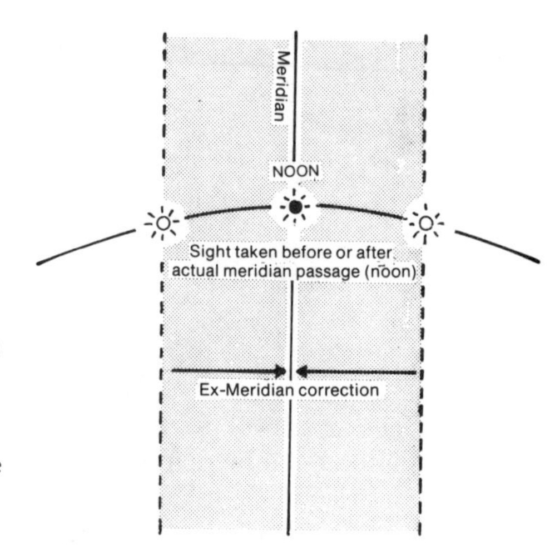

The ex-meridian correction is applied to the altitude taken before or after meridian passage in order to reduce this altitude to the actual noon meridian passage altitude.

which are indicated in Tables 29 and 30 of Vol. II of *The American Practical Navigator* (Bowditch). When there is some doubt about obtaining a sight right on meridian passage—as is often the case when there is a partial cloud cover—the limits should be calculated before hand. Within these limits an ex-meridian sight can be taken at any time, but obviously the closer to meridian passage the sight is taken, the smaller the correction to the altitude and the lower the risk of error.

The procedure for taking an ex-meridian sight of the sun is as follows:

1 Take a sextant altitude of the sun as close as possible to meridian passage (noon) and record the exact GMT.

2 Correct the sextant for normal sight errors and determine the interval, in minutes and seconds, between the time of the sight and meridian passage.

3 Obtain the declination from *The Nautical Almanac*.

4 Using lat and dec (same name or different name) as arguments, enter Table 29 and take out quantity 'a' (Altitude Factor).

5 With the interval between the sight and meridian passage (Meridian Angle) enter Table 30 across the top. Read off the factor which appears against 'a'.

6 This is the correction in minutes and decimals of a minute to be *added* to the ex-meridian altitude to convert it to the meridian altitude.

7 Complete the sight as for normal Latitude by meridian altitude sight.

The ex-meridian sight can be very useful when weather conditions make sight taking difficult.

Latitude by Polaris

The star Polaris is located almost exactly over the North Pole. This means that when viewed from anywhere in the world, it is bearing virtually due north and therefore making a permanent meridian passage. For this reason a latitude can be obtained from Polaris at any time the horizon is clear enough. This creates an interesting phenomenon in the 'latitude by meridian passage' formula. The best way to illustrate this is to use an example. Assume the altitude of Polaris to be 50°. Then, working it as a meridian passage, the following result would be obtained:

	90°00′ −	
true alt Polaris	50°00′	
ZX	40°00′	(south of star)
Dec of Polaris	90°00′	N
Latitude	50°00′	

From this example it is obvious that the latitude of the observer is the same as the altitude of Polaris and indeed this is the case. Correction of the sextant, plus minor corrections which are the result of the star being fractionally out of alignment with the Pole, provides an accurate latitude read literally straight from the sextant.

Pole Star tables

In the back of *The Nautical Almanac*, are the Pole Star Tables. There are three tables, each providing a correction to be applied to the true altitude of Polaris. These corrections are named a_0, a_1, a_2. Using these corrections, the formula for finding the latitude by the Pole star is as follows:

Latitude = Apparent Altitude - 1° + a_0 + a_1 + a_2

Polaris computation

The following is the procedure for resolving the Polaris sight:

1 From the daily pages of *The Nautical Almanac* determine the GHA of Aries for the time of observation.

2 By applying the longitude (east plus, west minus) convert the GHA to LHA of Aries.

3 Enter the first Pole Star Table and take out the correction a_0 for the LHA of Aries as closely as possible.

4 Find the correction a_1 in the same column against the latitude listed down the side.

274 POLARIS (POLE STAR) TABLES, 1978
FOR DETERMINING LATITUDE FROM SEXTANT ALTITUDE AND FOR AZIMUTH

L.H.A. ARIES	0°–9°	10°–19°	20°–29°	30°–39°	40°–49°	50°–59°	60°–69°	70°–79°	80°–89°	90°–99°	100°–109°	110°–119°
	a_0	a_0	a_0	a_0	a_0	a_0	a_0	a_0	a_0	a_0	a_0	a_0
0	0 16·9	0 12·8	0 10·1	0 08·9	0 09·2	0 11·1	0 14·4	0 19·1	0 25·1	0 32·0	0 39·8	0 48·2
1	16·4	12·4	09·9	08·8	09·3	11·4	14·8	19·7	25·7	32·8	40·6	49·0
2	16·0	12·1	09·7	08·8	09·5	11·6	15·3	20·2	26·4	33·5	41·5	49·9
3	15·5	11·8	09·5	08·8	09·6	11·9	15·7	20·8	27·1	34·3	42·3	50·7
4	15·1	11·5	09·4	08·8	09·8	12·3	16·2	21·4	27·8	35·1	43·1	51·6
5	0 14·7	0 11·2	0 09·3	0 08·8	0 10·0	0 12·6	0 16·6	0 22·0	0 28·4	0 35·8	0 43·9	0 52·5
6	14·2	11·0	09·2	08·9	10·2	12·9	17·1	22·6	29·2	36·6	44·8	53·3
7	13·9	10·7	09·1	08·9	10·4	13·3	17·6	23·2	29·9	37·4	45·6	54·2
8	13·5	10·5	09·0	09·0	10·6	13·6	18·1	23·8	30·6	38·2	46·5	55·1
9	13·1	10·3	08·9	09·1	10·8	14·0	18·6	24·4	31·3	39·0	47·3	55·9
10	0 12·8	0 10·1	0 08·9	0 09·2	0 11·1	0 14·4	0 19·1	0 25·1	0 32·0	0 39·8	0 48·2	0 56·8
Lat.	a_1	a_1	a_1	a_1	a_1	a_1	a_1	a_1	a_1	a_1	a_1	a_1
0	0·5	0·6	0·6	0·6	0·6	0·5	0·5	0·4	0·3	0·3	0·2	0·2
10	·5	·6	·6	·6	·6	·5	·5	·4	·4	·3	·3	·2
20	·5	·6	·6	·6	·6	·6	·5	·5	·4	·4	·3	·3
30	·6	·6	·6	·6	·6	·6	·5	·5	·5	·4	·4	·4
40	0·6	0·6	0·6	0·6	0·6	0·6	0·6	0·5	0·5	0·5	0·5	0·5
45	·6	·6	·6	·6	·6	·6	·6	·6	·6	·5	·5	·5
50	·6	·6	·6	·6	·6	·6	·6	·6	·6	·6	·6	·6
55	·6	·6	·6	·6	·6	·6	·6	·6	·7	·7	·7	·7
60	·6	·6	·6	·6	·6	·6	·7	·7	·7	·8	·8	·8
62	0·7	0·6	0·6	0·6	0·6	0·6	0·7	0·7	0·8	0·8	0·8	0·8
64	·7	·6	·6	·6	·6	·6	·7	·7	·8	·8	·9	0·9
66	·7	·6	·6	·6	·6	·7	·7	·8	·8	0·9	0·9	1·0
68	0·7	0·6	0·6	0·6	0·6	0·7	0·7	0·8	0·9	1·0	1·0	1·1
Month	a_2	a_2	a_2	a_2	a_2	a_2	a_2	a_2	a_2	a_2	a_2	a_2
Jan.	0·7	0·7	0·7	0·7	0·7	0·7	0·7	0·7	0·7	0·7	0·7	0·6
Feb.	·6	·6	·7	·7	·7	·8	·8	·8	·8	·8	·8	·8
Mar.	·5	·5	·6	·6	·7	·7	·8	·8	·8	·9	·9	·9
Apr.	0·3	0·4	0·4	0·5	0·6	0·6	0·7	0·7	0·8	0·9	0·9	0·9
May	·2	·2	·3	·3	·4	·5	·5	·6	·7	·8	·8	·9
June	·2	·2	·2	·2	·3	·3	·4	·5	·5	·6	·7	·8
July	0·2	0·2	0·2	0·2	0·2	0·2	0·3	0·3	0·4	0·5	0·5	0·6
Aug.	·3	·3	·3	·2	·2	·2	·2	·3	·3	·3	·4	·4
Sept.	·5	·5	·4	·4	·3	·3	·3	·3	·3	·3	·3	·3
Oct.	0·7	0·6	0·6	0·5	0·5	0·4	0·4	0·3	0·3	0·3	0·3	0·3
Nov.	0·9	·8	·8	·7	·6	·6	·5	·4	·4	·3	·3	·3
Dec.	1·0	0·9	0·9	0·9	0·8	0·7	0·7	0·6	0·5	0·5	0·4	0·4
Lat.	AZIMUTH											
0	0·4	0·3	0·1	0·0	359·8	359·7	359·6	359·4	359·3	359·3	359·2	359·2
20	0·4	0·3	0·1	0·0	359·8	359·7	359·5	359·4	359·3	359·2	359·2	359·1
40	0·5	0·3	0·1	0·0	359·8	359·6	359·4	359·3	359·1	359·0	359·0	358·9
50	0·6	0·4	0·2	359·9	359·7	359·5	359·3	359·1	359·0	358·8	358·8	358·7
55	0·7	0·5	0·2	359·9	359·7	359·4	359·2	359·0	358·8	358·7	358·6	358·6
60	0·8	0·5	0·2	359·9	359·6	359·4	359·1	358·9	358·7	358·5	358·4	358·3
65	0·9	0·6	0·3	359·9	359·6	359·2	358·9	358·6	358·4	358·2	358·1	358·0

Latitude = Apparent altitude (corrected for refraction) $- 1° + a_0 + a_1 + a_2$

The table is entered with L.H.A. Aries to determine the column to be used; each column refers to a range of 10°. a_0 is taken, with mental interpolation, from the upper table with the units of L.H.A. Aries in degrees as argument; a_1, a_2 are taken, without interpolation, from the second and third tables with arguments latitude and month respectively. a_0, a_1, a_2 are always positive. The final table gives the azimuth of *Polaris*.

Extract from the Pole Star Tables in *The Nautical Almanac*.

5 Find the correction a2 in the same column against the month of the year listed down the side.
6 Correct the sextant altitude as for any star to obtain the apparent altitude and subtract 1°.
7 To the result add the sum of the corrections a0, + a1 + a2. The result is the latitude.

Compass checking at sea

It is important to check the master compass at least once a day when making a passage. However, since there are no land objects in sight when making an ocean passage, heavenly bodies must be used to determine if the compass has an error. To minimise the amount of work, I find it best to check the compass by sun at the same time as taking the morning sight. However, there is no reason why any star, planet or even the moon cannot be used at any time in exactly the same way for compass checking, and the only reason for combining the morning sight and the compass check is that they use the same computation, and therefore an extra sight reduction is avoided.

The heavenly body used should be low in the sky for preference to obtain a more accurate bearing. The procedure, using the sun as an example, is as follows:

1 With an azimuth mirror or some similar sighting device, take a bearing of the sun with the compass.
2 Correct this bearing for variation and deviation to obtain a true bearing of the sun.
3 Note carefully the GMT at which this bearing was taken, and using *The Nautical Almanac* and the Sight Reduction Form, obtain the Tab LHA, the Tab Dec and the Dec Inc of the sun at that GMT.
4 Obtain an assumed latitude from the DR position.
5 Enter the sight reduction tables as normal, but take out only the value Z.
6 Apply 180° as indicated to obtain the value Zn.
7 Zn is the true azimuth of the sun at the time of the observation and should be the same as the true bearing obtained from the compass.
8 Any discrepancy between these two azimuths indicates an error in the compass.

Sunset and sunrise

The daily pages of *The Nautical Almanac* have, on their right-hand side, a list of times of sunrise and sunset, moonrise and moonset. Although, like all timing in *The Nautical Almanac*, these values are in GMT, they indicate the GMT for Greenwich meridian (0°) and therefore can be taken as the LMT (ship's time) for any other meridian. Thus they can be read out as local time providing the boat's clock is corrected for her longitude. The times of rising and setting are listed according to latitude.

Twilights

The times given for twilight, which are also LMT, are tabulated for nautical and civil twilights. They indicate the commencement of dawn, prior to sunrise, and the time of darkness after sunset and are listed because of their usefulness in determining the amount of time available for taking sextant observations of stars. Like sunset and sunrise, they are tabulated according to latitude.

The difference between nautical and civil twilight is a question of degree and darkness and at this stage you would be best advised to use civil twilight which usually allows for good star observations with a clear, sharp horizon.

Lat.	Twilight Naut.	Twilight Civil	Sunrise	Moonrise 2	Moonrise 3	Moonrise 4	Moonrise 5
°	h m	h m	h m	h m	h m	h m	h m
N 72	□	□	□	25 13	01 13	02 54	04 33
N 70	////	////	01 40	00 33	01 54	03 22	04 51
68	////	////	02 23	01 09	02 22	03 42	05 05
66	////	00 47	02 51	01 35	02 43	03 58	05 16
64	////	01 48	03 13	01 55	03 00	04 11	05 25
62	////	02 21	03 30	02 11	03 14	04 22	05 33
60	00 43	02 45	03 44	02 24	03 25	04 31	05 40
N 58	01 38	03 03	03 56	02 36	03 35	04 39	05 46
56	02 08	03 19	04 07	02 46	03 44	04 47	05 52
54	02 30	03 32	04 16	02 54	03 52	04 53	05 57
52	02 48	03 43	04 24	03 02	03 59	04 59	06 01
50	03 03	03 53	04 31	03 09	04 05	05 04	06 05
45	03 31	04 13	04 47	03 24	04 19	05 15	06 13
N 40	03 52	04 29	05 00	03 37	04 30	05 25	06 20
35	04 09	04 43	05 10	03 48	04 39	05 32	06 27
30	04 23	04 54	05 20	03 57	04 48	05 39	06 32
20	04 45	05 12	05 36	04 13	05 02	05 51	06 41
N 10	05 02	05 28	05 50	04 27	05 14	06 02	06 49
0	05 16	05 41	06 03	04 40	05 26	06 12	06 57
S 10	05 28	05 53	06 15	04 53	05 38	06 22	07 04
20	05 39	06 06	06 29	05 07	05 50	06 32	07 12
30	05 50	06 19	06 44	05 23	06 04	06 44	07 21
35	05 56	06 26	06 53	05 32	06 13	06 51	07 27
40	06 02	06 34	07 03	05 42	06 22	06 59	07 33
45	06 08	06 43	07 15	05 55	06 33	07 08	07 39
S 50	06 15	06 54	07 29	06 10	06 46	07 19	07 48
52	06 18	06 59	07 36	06 17	06 52	07 24	07 52
54	06 21	07 04	07 43	06 24	06 59	07 29	07 56
56	06 24	07 10	07 51	06 33	07 07	07 35	08 00
58	06 28	07 16	08 00	06 43	07 15	07 42	08 05
S 60	06 32	07 23	08 11	06 54	07 25	07 50	08 11

Lat.	Sunset	Twilight Civil	Twilight Naut.	Moonset 2	Moonset 3	Moonset 4	Moonset 5
°	h m	h m	h m	h m	h m	h m	h m
N 72	□	□	□	21 24	21 18	21 12	21 07
N 70	22 26	////	////	20 42	20 50	20 53	20 55
68	21 45	////	////	20 14	20 28	20 38	20 45
66	21 18	23 13	////	19 52	20 12	20 26	20 37
64	20 57	22 19	////	19 35	19 58	20 16	20 30
62	20 40	21 48	////	19 21	19 46	20 07	20 24
60	20 26	21 25	23 19	19 09	19 36	19 59	20 19
N 58	20 15	21 07	22 30	18 58	19 28	19 53	20 14
56	20 04	20 52	22 01	18 49	19 20	19 47	20 10
54	19 55	20 39	21 40	18 41	19 13	19 41	20 07
52	19 47	20 28	21 22	18 34	19 07	19 37	20 03
50	19 40	20 18	21 08	18 27	19 02	19 32	20 00
45	19 25	19 58	20 40	18 13	18 50	19 23	19 53
N 40	19 12	19 42	20 19	18 02	18 40	19 15	19 48
35	19 01	19 29	20 02	17 52	18 31	19 08	19 43
30	18 52	19 18	19 49	17 43	18 24	19 02	19 39
20	18 36	19 00	19 27	17 28	18 11	18 51	19 31
N 10	18 22	18 45	19 10	17 15	17 59	18 42	19 24
0	18 10	18 31	18 56	17 02	17 48	18 34	19 18
S 10	17 57	18 19	18 44	16 50	17 37	18 25	19 12
20	17 44	18 07	18 33	16 36	17 26	18 15	19 05
30	17 28	17 53	18 22	16 21	17 13	18 05	18 57
35	17 20	17 46	18 17	16 12	17 05	17 58	18 53
40	17 10	17 38	18 11	16 02	16 56	17 51	18 48
45	16 58	17 29	18 05	15 50	16 46	17 43	18 42
S 50	16 44	17 19	17 58	15 36	16 33	17 33	18 35
52	16 37	17 14	17 55	15 29	16 27	17 28	18 31
54	16 30	17 09	17 52	15 21	16 21	17 23	18 28
56	16 22	17 03	17 49	15 13	16 14	17 18	18 24
58	16 13	16 57	17 45	15 03	16 05	17 11	18 19
S 60	16 02	16 50	17 41	14 52	15 56	17 04	18 14

Day	SUN Eqn. of Time 00ʰ	SUN Eqn. of Time 12ʰ	SUN Mer. Pass.	MOON Mer. Pass. Upper	MOON Mer. Pass. Lower	MOON Age	MOON Phase
	m s	m s	h m	h m	h m	d	
2	06 15	06 13	12 06	10 51	23 14	28	●
3	06 11	06 09	12 06	11 37	24 00	29	
4	06 06	06 04	12 06	12 23	00 00	00	

The sunrise and twilight tables from *The Nautical Almanac*.

Checking the compass at sea is an important daily routine.

Chapter 10 Star Identification

Most newcomers to navigation find identification of heavenly bodies one of the most difficult aspects of taking a star sight. Other bodies, such as sun, moon and to a certain extent the planets, are obvious and easily identified. The stars, however, since there are so many of them, tend to be confusing and difficult to isolate one from the other. Particularly is this the case in constellations such as *Orion* where four or five major stars with almost the same brilliance are grouped together. The confusion is compounded when the constellation swings across the heavens seeming to change its shape and direction. The stars nearest the pole appear to move slower than those near the equator.

This 'wheeling' effect is like a line of soldiers marching line abreast and wheeling around a corner. The soldier on the inside of the line will remain virtually stationary and mark time while the soldier on the outside of the line will have to almost run to keep up. Stars which are located close to the pole have very little apparent movement since in order to describe their daily 360° circle around the world, they move only relatively little. Stars on the equator, by contrast, have to travel right around the world's girth in the same amount of time (twenty-four hours) and therefore appear to move much faster.

Like all other heavenly bodies, stars rise in the east and set in the west, although those closest to the pole, as described, make a very small orbit around the pole and therefore will be above the horizon for the entire twenty-four hour period when seen by an observer in moderately high latitudes. Polaris, the classic North Pole Star, for example, appears to hardly move from dusk to dawn and from day to day. Other stars, however, particularly those with low declinations, carry out the normal rising and setting phenomenon which applies to other heavenly bodies.

The constellations

Without doubt, the easiest method of star identification is to recognise the constellations or groups formed by navigational stars of first magnitude. There are many ways of identifying stars, and these will be dealt with later in this section, but for the beginner, identification through constellations is particularly important. Since he will be using these stars for his daily star sights, it is imperative that he gets to know these celestial signposts which indicate the stars he needs. The grouping of the constellations is such that as a general rule one leads easily to another and thus identification of one prominent star or constellation can, as it were, set off a chain reaction which will lead to identification of most of the stars in the area, if not the whole night sky.

Stars are taken at dawn and dusk in the period known as twilight and therefore two different sets of stars—with possibly a common polar star—will be used. There is plenty of time to identify the dawn stars before the sky lightens to the point where the horizon is sufficiently sharp for sights to be taken. At dusk, however, by the time the stars are sufficiently bright for visual identification, the horizon is too dark for sights. Instant identification is necessary when the first faint twinkle of a star is seen and this is where the constellation pattern is probably the best, since it is the first magnitude stars that will appear early and offer a chance of identification before the horizon gets too dark. Once a voyage is under way, the stars remain in virtually the same position from night to night and, after the initial twilight star taking, identification on subsequent nights is much simplified.

The Plow or Big Dipper

Ursa Major, or the Big Dipper, as it is called, is the most important polar constellation in the Northern Hemisphere. It involves the use of 'pointers' which in this case point to the most useful navigational star of all—Polaris, the Pole Star. A reverse line through these pointers lead to another navigational star—Regulus, so this constellation serves a twofold purpose.

Orion

This is a constellation readily identified by its shape. The four stars (which virtually form a cross) have as a focal point between them the three small stars known as 'Orion's Belt'. This makes it a very distinctive constellation and one which, providing it is

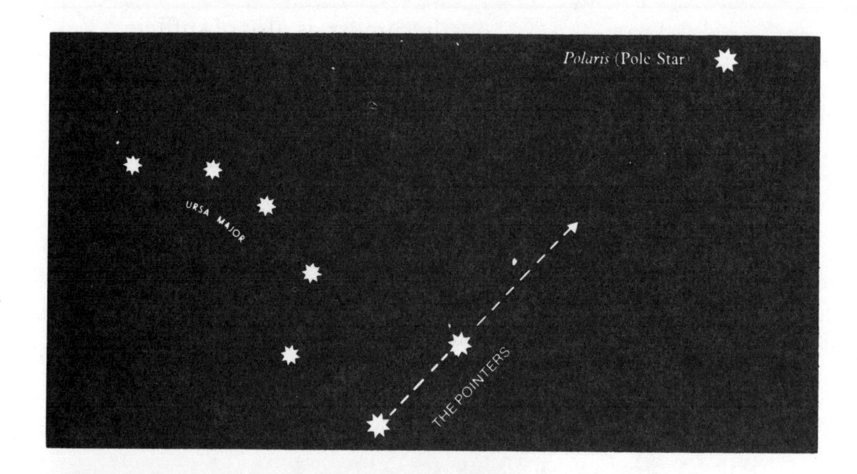

The Plow or Big Dipper constellation with Polaris.

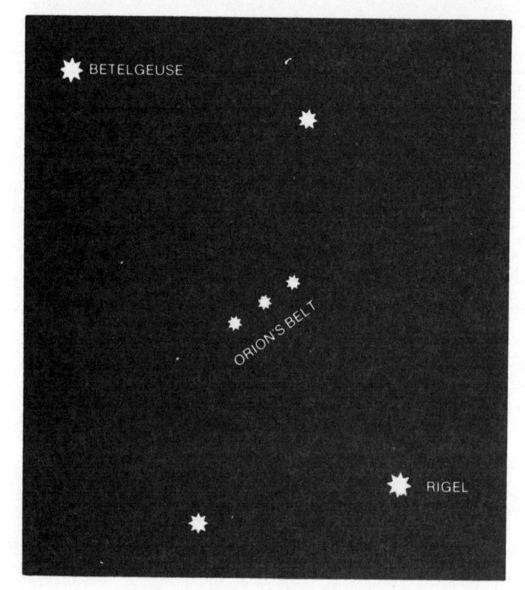

The constellation of Orion.

above the horizon, is readily visible. Since it is more an equatorial constellation than a polar group, Orion can be easily seen in most parts of the world.

Because they are grouped closely together, the four stars of the constellation cannot be used for the same navigational sight. One, or perhaps two, well-spaced stars may be used and thus identification of each one is important. Rigel is the southernmost of the constellation and without doubt the best star for navigational purposes; Betelgeuse, the northernmost, is also of sufficient magnitude to use for sights. The remaining stars are less brilliant and should not be used. Perhaps the greatest benefit that Orion offers the navigator lies not in its own stars, but in two immediate constellations to which it leads.

Canis Major (Great Dog)

Follow the line of the three stars in Orion's Belt to the south and you will find one of the most brilliant stars in the sky—Sirius. This magnificient star is quite unique, twinkling brilliantly with all the colors of the spectrum. Sirius is unmistakable when compared with other stars in its vicinity and is the only star of any prominence in the Great Dog constellation.

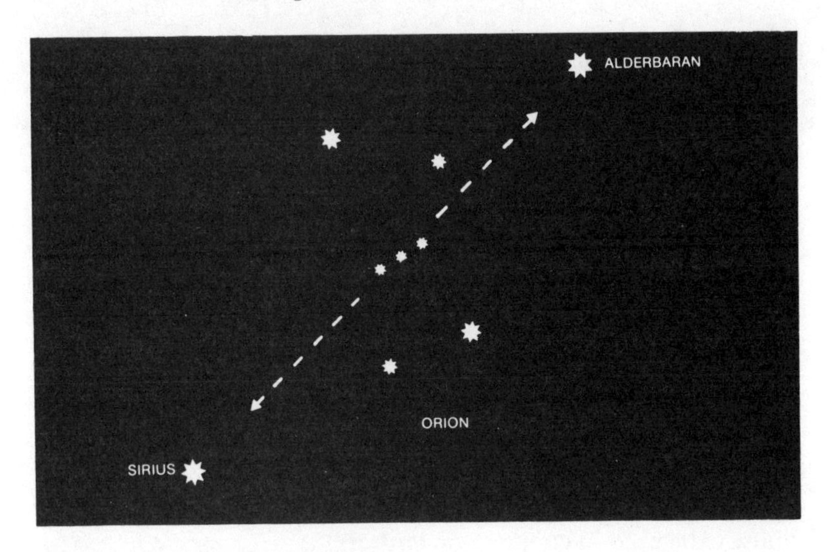

The leads to Sirius and Aldebaran.

Taurus (The Bull)

Like Canis Major, Taurus has only one brilliant star—Aldebaran, As with Sirius, Aldebaran is found following the three stars of Orion's Belt, this time to the northwards. Again, Aldebaran and Sirius have distinctive characteristics, while the latter radiates brilliant twinkling colors, Aldebaran is a dullish, yellow-colored star. There are no other stars in this constellation useful for star sights.

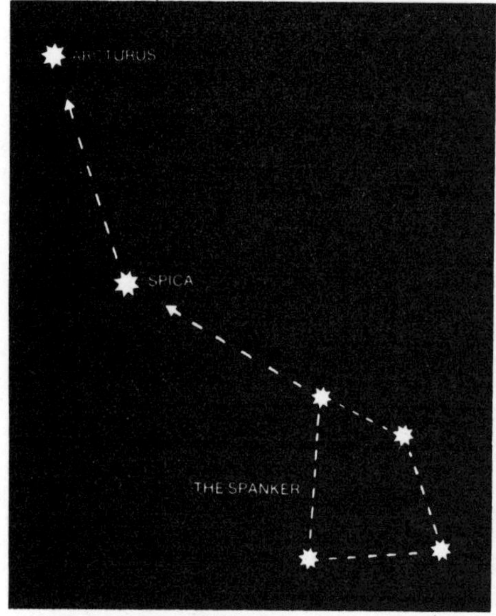

Spica's Spanker.

The Scorpion constellation.

The Spanker

This constellation is relatively small but again has quite a distinctive shape. It is high in the sky when viewed from either hemisphere since its declination is close to the equator. The Spanker is in effect an uneven rectangle shaped like a gaff sail or, as they were termed in the days of windjammers, a spanker. The gaff or head of the sail leads directly to Spica, a brilliant first magnitude star widely used for navigation. Northwards from Spica is Arcturus, another first magnitude star but one which is mostly too close to Spica to be used in conjunction with it.

The Scorpion

To the east of The Spanker constellation and slightly to the south of it lies The Scorpion constellation which has one prominent navigational star named Antares. Like Aldebaran, Antares is more yellow than white and stands out for this very reason, but it can also be identified as the apex of three stars in a very shallow triangle not unlike Orion's Belt.

The Southern Cross constellation.

The Southern Cross

The most common and easily identified constellation in the Southern Hemisphere is the Southern Cross. While it is not exactly a polar constellation, it has sufficiently high declination to ensure that it remains above the horizon for most of the night in high southern latitudes. It is of a very distinctive form, as illustrated, and can be found by facing roughly south and looking for the cross formation of the four stars. Only one of these—Acrux—is of any prominence, the others being of a relatively lesser magnitude. To the east of the cross itself are the pointers which are roughly in line with the arms of a cross. The outermost pointer, a prominent and useful first magnitude star called Rigil Kentaurus, is a good starting point to lead to other stars of use in navigation.

Follow the pointers from Rigil Kentaurus through the Southern Cross westwards to another similar constellation—although less bright and more scattered—which is known as the 'False Cross' constellation. There are no stars of use to navigation in this constellation but continuing slightly westwards again is one of the most brilliant stars in the southern sky—Canopus. This is an ideal star to use with Rigil Kent. Continuing again towards the west beyond Canopus is Achernar which is another excellent navigational star. Thus, from the Southern Cross constellation and its pointer Rigil Kent, three brilliant navigational stars can easily be tracked across the southern sky and identified—providing, of course, that they are above the horizon.

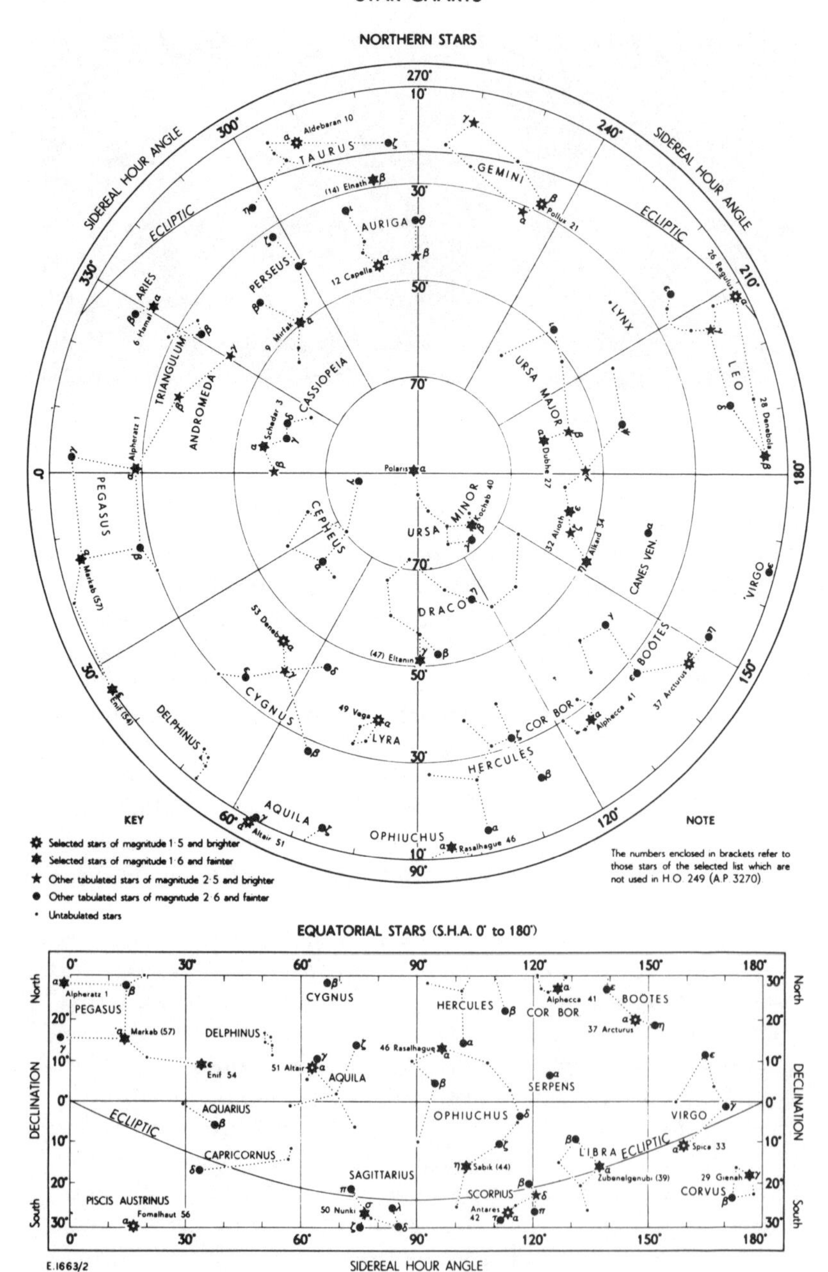

A typical star chart from *The Nautical Almanac.*

Star charts

There are many different forms of star or astro charts available on the market, each of which employs its own individual technique for star identification. The simplest merely illustrate the constellations and their relation to one another, with first magnitude stars marked prominently. Others are called star 'finders' since they not only identify the star but provide its azimuth and altitude at a predetermined time. I do not propose to deal in detail with these charts and star finders since they are too numerous and too individual. However, all come with step-by-step instructions on their use and it is more a question of personal preference as to which one helps make star identification easiest. There are star charts in the back of *The Nautical Almanac* for both Northern and Southern Hemispheres as well as for equatorial stars.

Making a 'personal' star chart

One of the methods I have found most practical in teaching beginners how to identify stars prior to sight taking on a voyage is the compilation of your own 'personal' star chart. As mentioned there is no great problem with dawn stars since you will have all night to sit down and identify them, and for that matter take practice sights with the dark horizon in preparation for the dawn shots. With dusk stars, however, there is no opportunity to identify either the stars or their constellations. By the time you get round to identification, the twilight has faded and the horizon is indistinct.

Lets assume that you are setting out on an ocean passage tomorrow morning and are anxious to know which stars you will be using and where in the sky they will be found. Then the following procedure will provide not only a star chart of the stars you intend to use but also a rough guide to their azimuth and altitude:

1 On the night before you leave take yourself down to an ocean beach—or even better take the boat out into open water—armed with a sextant and hand-bearing compass.
2 Draw on a piece of paper an outline of your boat heading on the course that she will follow on the first night at sea. As soon as possible after twilight has disappeared, face along the beach (or head your boat) in this direction and by means of a star chart or by studying the previous paragraph in this section, identify a prominent constellation—say The Big Dipper.

The basic chart.

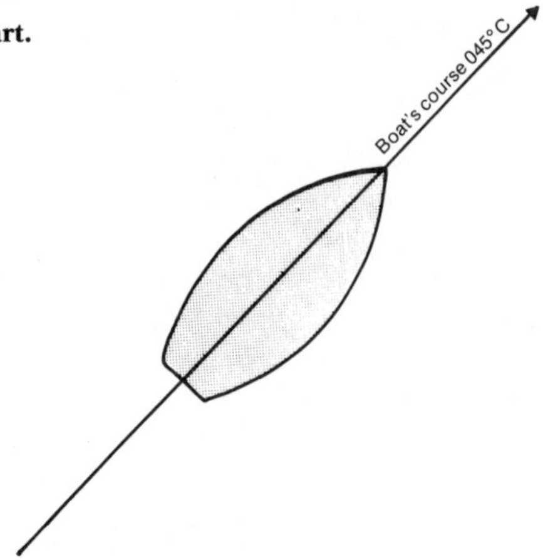

3 Take a sextant altitude of the most important star—Polaris—using the now-dark horizon as accurately as possible.
4 With the hand-bearing compass obtain a bearing of the same star.
5 On your star chart, lay off the bearing from a central position and mark against it the compass bearing and the sextant altitude of the star.
6 Repeat this procedure with all other visible stars that are well spaced for sight taking. Ideally these should be not less than 30° between bearings.
7 When completed, you will have a personalised star chart of the stars that will be available for your dusk sights the following night. In addition you will have the approximate azimuth and altitude of each star.

The advantages of this system are that on the following night, as twilight approaches, you can pre-set the sextant to the approximate altitude of each star and sight towards the horizon along its azimuth long before the star becomes visible to the naked eye. The magnification of the sextant telescope will show up the star before it is visible in the sky. The azimuth will indicate the identity of the star so that all that will be required is fine adjustment to the micrometer screw to bring the star right down to the horizon to obtain an accurate altitude. The horizon will still be bright and clear. The choice of stars will enable you to pick four or more that are clear of cloud or where the horizon is brightest (to the west) thus offering every chance of a good star sight without needing to see the full sky and the constellations themselves.

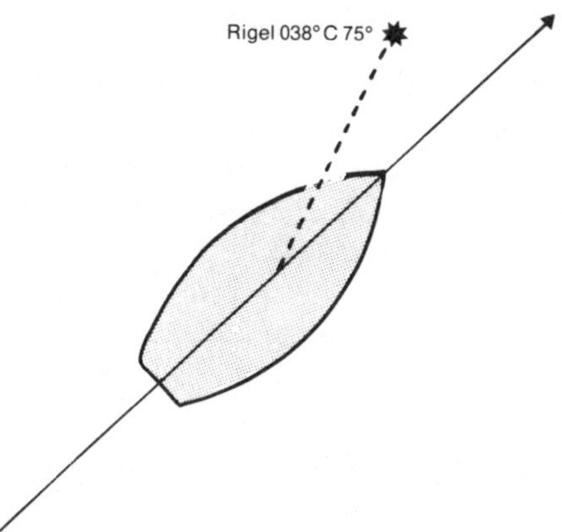

Laying off the first approximate altitude and azimuth.

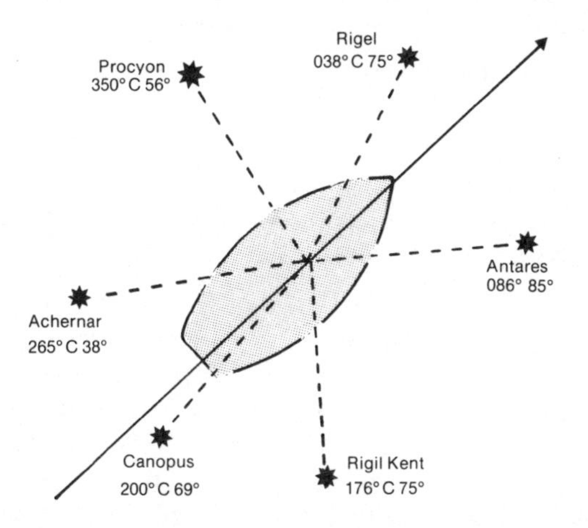

The completed 'personal' star chart. The illustrated stars are chosen at random and do not necessarily relate to the night sky in any particular location.

This system, as described below, is widely practised by professional navigators, although they would not bother making a personalised chart, but simply pre-compute the altitude and azimuth of each star from *The Nautical Almanac*. On each of the following nights you will be able to pre-set your sextant more and more accurately as you become more familiar with the stars. There is little change in the altitude or azimuth of each star from night to night.

Pre-computation

If the time of star sights is known within a reasonable margin of error, then the altitude and azimuth of each star can be pre-computed to within a matter of minutes of arc. The sextant can then be pre-set to use the technique described above for sighting the star in the sextant telescope before it is visible to the naked eye. Since twilight is listed in *The Nautical Almanac* on the right-hand side of the daily pages, the approximate time of star sights can be found simply by reference to this table. Star sights are usually taken close to the end of civil twilight. Having obtained the GMT of the approximate time of shooting the star sight, the procedure for pre-computing the altitude and azimuth of each star is as follows:

1 Establish your DR position for the time of sights. Using Section 3 of my Sight Reduction Form, work up the required data to use with the sight reduction tables as indicated. This is simply a normal sight procedure using the GMT you have estimated for sight taking.
2 Enter the sight reduction tables and obtain Hc and Z.
3 Convert Z to Zn as for normal sights.
4 Hc then represents the approximate altitude of the star and Zn its azimuth.

It follows that since the GMT of sight taking is only approximate, as is the DR position, the altitude and azimuth will be also be only approximate. However, within quite a reasonable range, the altitude found in this way will ensure that the star appears in the sextant telescope even though it may not be close to the horizon. Similarly, the azimuth will be as close as necessary and if the star does not appear immediately a modest sweep of the horizon on either side will almost certainly locate it.

As mentioned, this is the system used by professional navigators the world over for pre-computing altitudes and azimuths and pre-setting their sextants. It enables them to spot each star and obtain a sight before the horizon has even begun to darken.